International Research on the Impact of Accountability Systems

Teacher Education Yearbook XV

EDITED BY LOUISE F. DERETCHIN
AND CHERYL J. CRAIG

Published in partnership with the
Association of Teacher Educators
ROWMAN & LITTLEFIELD EDUCATION
Lanham, Maryland • Toronto • Plymouth, UK
2007

Published in partnership with the
Association of Teacher Educators

Published in the United States of America
by Rowman & Littlefield Education
A Division of Rowman & Littlefield Publishers, Inc.
A wholly owned subsidiary of The Rowman & Littlefield Publishing Group, Inc.
4501 Forbes Boulevard, Suite 200, Lanham, Maryland 20706
www.rowmaneducation.com

Estover Road
Plymouth PL6 7PY
United Kingdom

Copyright © 2007 by Louise F. Deretchin and Cheryl J. Craig

All rights reserved. No part of this publication may be reproduced,
stored in a retrieval system, or transmitted in any form or by any
means, electronic, mechanical, photocopying, recording, or otherwise,
without the prior permission of the publisher.

British Library Cataloguing in Publication Information Available

Library of Congress Control Number: 2006930214

ISBN-13: 978-1-57886-546-8 (cloth, alkaline paper)
ISBN-10: 1-57886-546-8 (cloth, alkaline paper)
ISBN-13: 978-1-57886-547-5 (pbk., alkaline paper)
ISBN-10: 1-57886-547-6 (pbk., alkaline paper)

∞ ™ The paper used in this publication meets the minimum requirements of
American National Standard for Information Sciences—Permanence of
Paper for Printed Library Materials, ANSI/NISO Z39.48-1992.
Manufactured in the United States of America.

Teacher Education Yearbook XV

EDITORS

Louise F. Deretchin, Houston A+ Challenge
Cheryl J. Craig, University of Houston

EDITORIAL ADVISORY BOARD

Freddie Bowles, University of Arkansas
Tammie Brown, Middle Tennessee State University
Caroline Crawford, University of Houston–Clear Lake
Susan Graff, Barry University
Linda Houser, Indiana University, Purdue University–Indianapolis
Rudy P. Mattai, State University of New York at Buffalo
Frances Van Tassel, University of North Texas
David M. Byrd, University of Rhode Island
Julie Rainer Dangel, Georgia State University
John McIntyre, Southern Illinois University at Carbondale
Sandra J. Odell, University of Nevada at Las Vegas
Mary John O'Hair, University of Oklahoma

EDITORIAL SUPPORT BOARD

Blake R. Bickham, University of Houston
Neil J. Liss, Willamette University
Kyongmoon Park, University of Houston
Rita P. Poimbeauf, University of Houston
Mark L. Seaman, University of Houston

EXECUTIVE DIRECTOR

David Ritchey, Association of Teacher Educators, Manassas Park, Virginia

Contents

List of Illustrations	ix
Foreword	xi
Jane McCarthy	
Introduction	xiii
Louise F. Deretchin and Cheryl J. Craig	

DIVISION 1: PHILOSOPHY, HISTORY, AND DESIGN OF ACCOUNTABILITY SYSTEMS

Overview and Framework	3
Michaelann Kelley, Donna Reid, Gayle Curtis, Ron Venable, P. Tim Martindell, and Allison Hamacher	
1 The Uses and Abuses of Measurement	7
Margaret J. Wheatley and Myron Rogers	
2 Teachers' Self-Understanding in Times of Performativity	13
Geert Kelchtermans	
3 Standards and Accountability	31
David J. Flinders	
Summary and Implications	43
Michaelann Kelley, Donna Reid, Gayle Curtis, Ron Venable, P. Tim Martindell, and Allison Hamacher	

DIVISION 2: IMPACT AND CONSEQUENCES OF ACCOUNTABILITY SYSTEMS

Overview and Framework	47
Susan McCormack, Denise McDonald, Tirupalavanam G. Ganesh, and Andrea S. Foster	

4	Accountability Systems and Program Evaluation *Stephen Fletcher, Michael Strong, and Anthony Villar*	53
5	Revisiting the Impact of High-Stakes Testing on Student Outcomes from an International Perspective *Jaekyung Lee*	65
6	Broken Promises, High Stakes, and Consequences for Native Americans *Beverly J. Klug*	83
7	Accountability Systems' Narrowing Effect on Curriculum in the United States *Jennifer L. Snow-Gerono and Cheryl A. Franklin*	97
8	The Impact of Standardized Testing on Teachers' Pedagogy in Three 2nd-Grade Classrooms of Varied Socio-Economic Settings *Annapurna Ganesh*	113

Summary and Implications 129
Susan McCormack, Denise McDonald, Tirupalavanam G. Ganesh, and Andrea S. Foster

DIVISION 3: PERCEPTIONS AND PERSPECTIVES OF ACCOUNTABILITY SYSTEMS

	Overview and Framework *Michele Kahn, Mimi Miyoung Lee, Carrie Markello, Heidi C. Mullins, and Annapurna Ganesh*	135
9	No Teacher Educator Left Behind *Clare Kosnik*	139
10	Critique through Visual Data *Tirupalavanam G. Ganesh*	157
11	No Child Left Behind and Accountability through a Democratic Lens *Cindy Finnell-Gudwien*	169

Summary and Implications 181
Michele Kahn, Mimi Miyoung Lee, Carrie Markello, Heidi C. Mullins, and Annapurna Ganesh

DIVISION 4: UNDERPINNINGS OF POWERFUL ACCOUNTABILITY SYSTEMS

	Overview and Framework *Neil J. Liss*	185
12	The Antecedents of Success *Eila Estola, Anneli Lauriala, Säde-Pirkko Nissilä, and Leena Syrjälä*	189

13	Creating a Learning Space for Educators *Ora W. Y. Kwo*	207
14	Lessons from Te Kotahitanga for Teacher Education *Russell Bishop*	225

Summary and Implications 241
Neil J. Liss

Afterword 247
Cheryl J. Craig and Louise F. Deretchin

Illustrations

Tables

1.1	Some important distinctions between feedback and measurement.	10
4.1	2002 demographic statistics for three districts using mentor-based induction.	56
4.2	Characteristics of classes taught by novice teachers.	57
4.3	Characteristics of the mentor-based induction programs for three districts.	57
4.4	Models of student achievement using student and class variables.	60
5.1	High school exit exams and college entrance exams in Korea, Japan, England, and the United States.	70
5.2	Results of multiple regression of TIMSS 8th-grade math student outcomes on predictors including high-stakes testing (HST), educational aspiration, East Asia, and GNP.	77

Figures

5.1	National average mathematics scores of 1995 TIMSS 8th graders (black bars for countries with HST).	72
5.2	Percentages of 1995 TIMSS 8th graders who agree or strongly agree that they usually do well in mathematics (black bars for countries with HST).	73
5.3	Percentages of 1995 TIMSS 8th graders who report that they like mathematics or like math a lot (black bars for countries with HST).	74
5.4	Percentages of 1995 TIMSS 8th graders who aspire to finish secondary school or college/university (black bars for countries with HST).	76

10.1	Powerless.	162
10.2	Modern Sisyphus.	164
10.3	"No creative teaching."	166
13.1	Conceptualization of teachers' sustainability in adaptation to changes.	213
13.2	Tension between accountability agendas and critical focus.	222

Foreword

Jane McCarthy
University of Nevada, Las Vegas

> Jane McCarthy, Ed.D., is interim dean and professor, College of Education, UNLV, and currently serves as president of the Association of Teacher Educators (2006–2007). She is also the director of the UNLV Center for Accelerated Schools, a national school reform effort that works with schools with high percentages of children in at-risk situations. Her writing and research are concentrated in the fields of teacher education, school reform, and working in American Indian schools. She has consulted internationally, spending extensive time in Germany and India working with schools and teachers.

The theme of my presidency for the Association of Teacher Educators is "Reinventing the Educational Landscape: Restoring Hope, Heart and Vision for Teachers, Learners, and Communities." I selected this theme because the traditional ways of schooling, pre-K through 12, and the traditional methods of teacher education no longer seem to be meeting the needs of today's school populations and families. I can remember asking my father, toward the end of his life, how many children in his elementary school did not learn to read. He looked at me very puzzled and said, "We all learned to read." He grew up in an inner city as a child of Irish immigrants and went to school with other first- and second-generation immigrant children. He was unusual in that he actually finished high school and some college. And yet, they all learned to read and write. I have seen correspondence he has had with his peers over the years, and all letters were well written, grammatically correct, and with few, if any, misspelled words.

How did the educational system of the early nineteen hundreds facilitate the successful mastery of basic skills for what was a generation of poor, immigrant children, many of whom spoke English as a second language? We really can only surmise at this point since standardized testing was not really developed until World War II. How could teachers, many of whom had only a normal school preparation program, be so successful? Again, any theories can only be suppositional. Why are today's students and teachers not as successful as those of the past? Or is the success of the past only a myth perpetuated by anecdotal evidence such as mine?

The current movement toward standards-based education and assessment and the whole accountability issue is taking place not just in this country but in others as well. Will this movement help renew the hope, heart, and vision of learners, teachers, and communities, or will it hinder the efforts educators make to facilitate the success of all students? This current volume seeks to address these issues and more surrounding the impact of accountability systems around the world. Are the data collected by such accountability endeavors actually used to improve student learning? Or do they in fact hinder student learning? Do teachers know what to do with the data that are collected? Are meaningful school improvement plans derived from the data or are the data used to "punish" what are considered "underperforming" students and "inadequate" schools?

Bernhardt (2003) suggests that schools can use data to analyze student performance and to revise learning strategies and school plans based on information collected. This requires, however, intensive professional development for teachers and administrators.

The authors and editors of this *Yearbook* look at the reality of what is happening across the world with regard to accountability. The book has four divisions that examine distinct aspects of the topic. The first looks at the philosophy, history, and design of the accountability movement. The second looks at the impact and consequences of accountability systems. Do they help, hinder, or are they neutral with regard to learner and teacher success? The third section looks at personal perceptions and perspectives of accountability on the part of teachers and teacher educators. The fourth division looks at underpinnings of powerful accountability systems, primarily from an international perspective. I won't spoil the ending by giving away conclusions the editors reach. Suffice it to say that Deretchin and Craig have put together an impressive array of research and reflections by prominent scholars and authors that address a wide array of issues. Their volume is a must-read for all educators whose lives are impacted by issues of accountability. (And that means all of us!) They have done the profession a great service in the production of this volume.

Reference

Bernhardt, V. L. (2003). *Using data to improve student learning in elementary schools.* Larchmont, NY: Eye on Education, Inc.

Introduction

Louise F. Deretchin
Houston A+ Challenge

Cheryl J. Craig
University of Houston

>Louise Deretchin, Ph.D., is director of higher education for the Houston A+ Challenge (formerly the Houston Annenberg Challenge). She has extensive experience in information technology and education. She is a cofounder of the Regional Faculty, whose purpose is to take a regional approach to directing the growth of educational systems. Her work focuses on creating collaborations among colleges, the business community, and school districts to improve teacher education, teaching, and learning. Deretchin is a fellow in the Association of Teacher Educators Leadership Academy and serves on the Texas Higher Education Board's P–16+ Council.

>Cheryl J. Craig, Ph.D., is professor and coordinator of the Teaching and Teacher Education Program at the University of Houston and the past president of the American Association of Teaching and Curriculum. Craig's most recent book is *Narrative Inquiries of School Reform: Storied Lives, Storied Landscapes, Storied Metaphors* (2003). Her research appears in such journals as the *American Educational Research Journal, Teaching and Teacher Education, Journal of Teacher Education*, and *Teachers College Record*.

When we, as coeditors, issued the call for prospective manuscripts for *International Research on the Impact of Accountability Systems: Teacher Education Yearbook XV*, we knew that accountability sat at the center of national and international debates concerning teaching and teacher education, but we had no idea how individual authors would respond to the seed of an idea nor the range of essays that the accountability theme would engender. Only as we (along with a panel of blind reviewers to whom we are grateful) poured over the manuscripts did the subtleties and complexities of how accountability policies play out in particular national and international settings become apparent, along with their global magnitude. Once the essays were received, we realized we had struck a deep chord with authors and that the accepted chapters approached the accountability theme in multi-dimensional ways. Four variations of the accountability theme quickly emerged around which this book cohered: Philosophy, History, and Design of Accountability Systems; Impact and Consequences of Accountability Systems; Perceptions and Perspectives of Accountability Systems; and Underpinnings of Powerful Accountability Systems.

For *Teacher Education Yearbook XV*, we decided to deviate from the tradition of inviting well-recognized scholars to provide commentaries on each of the topics. Instead, we turned it around and invited educators from schools and teacher preparation programs—the very people that are subjected to, and implementers of, accountability systems—to provide commentary on the scholarly works of well-recognized and up-and-coming researchers. By giving voice to these educators, an additional and valued perspective is gained.

In the *Yearbook*, each of the four topic areas is introduced and commentaries are provided by different sets of authors. Philosophy, History, and Design of Accountability Systems is introduced by members of the longest standing teacher research group of the Houston A+ Challenge (formerly the Houston Annenberg Challenge)—the Portfolio Group of Teachers. These teachers have consistently met to deliberate on their practices since 1998. The second and third sections, the Impact and Consequences and Perceptions, and Perspectives of Accountability Systems, are presented by different members of the Faculty Academy, an initiative of the Houston A+ Challenge school redesign movement. The purpose of the Faculty Academy is to increase school-university collaborative forms of research and to serve as members of a Regional Faculty in the Greater Houston area.

The final section, Underpinnings of Powerful Accountability Systems, is presented by a recent graduate of an education doctoral program. His analysis brings a deeply philosophical perspective to the overview and summary of the concluding division.

After reading the manuscripts selected for *Yearbook XV*, and experiencing the passion and intellect evident in the writings, we felt compelled to conclude this volume with an afterword that brings forth the interrelatedness and interweavings of the experiences, hopes, and fears expressed by chapter authors and respondents.

We hope that the following chapters which take a critical look at the use and misuse of accountability systems—the good, the bad, and the possible—will help schools, communities, and legislators make informed decisions on the design of accountability systems so that they may become welcomed enhancements to teaching and learning.

Division 1
PHILOSOPHY, HISTORY, AND DESIGN OF ACCOUNTABILITY SYSTEMS

Overview and Framework

Michaelann Kelley
Eisenhower High School

Donna Reid
Educational Consultant

Gayle Curtis
Hamilton Middle School

Ron Venable
Eisenhower High School

P. Tim Martindell
Houston A+ Challenge

Allison Hamacher
Drew Academy for Mathematics, Science, and the Arts

>Michaelann Kelley is a visual arts teacher at Eisenhower High School in the Aldine Independent School District. She has been teaching for the past 14 years. Michaelann is also a doctoral student at the University of Houston.

>Donna Reid is an educational consultant who supports the development of Critical Friends Groups in Houston schools. She taught history and English at the middle school level for eleven years.

>Gayle Curtis is the vanguard dean at Hamilton Middle School in Houston Independent School District. Her experience as a bilingual educator for 11 years includes teaching at the elementary, middle, and high school levels.

>Ron Venable has taught for 27 years. He works with students enrolled in regular and advanced placement studio art classes at Eisenhower High School. He has also been a member of the Eisenhower Critical Friends Group for nine years.

P. Tim Martindell works as a program coordinator for the Houston A+ Challenge. He is a doctoral student at the University of Houston and has 19 years of experience teaching middle and high school English in urban schools.

Allison Hamacher has been a teacher for 25 years. Having completed her master's degree in administration in 2004, she continues to teach with some administrative duties as well as to mentor new teachers.

Each author in this edited volume writes from authority. Those who present overviews and who frame the four sets of chapters also speak from authority. In the teacher research group, where we have consistently met to reflect on and account for our teaching practices, our authority arises from individual and collective experience springing from over 150 years of working in urban schools peopled by students from diverse socioeconomic and multicultural backgrounds. That authority of experience inevitably involves daily encounters in our classroom and school settings with the accountability phenomenon that forms the topic of this *Teacher Education Yearbook XV*. Living with accountability at the ground level necessarily informs this introduction to the chapters in the first division of this volume of international research into the impact of accountability systems.

The first three chapters of this book address similar aspects of the influences that accountability exerts on internal and external systems. In chapter 1, The Uses and Abuses of Measurement, Wheatley and Rogers make the important assertion that measurement is critical, but only to the extent that it tells us what we need to know in order to move forward in more instructive ways. For Wheatley and Rogers, measurement is one of a full range of tools, a means that brings about additional insights and spurs further action, rather than an end unto itself. The authors write that "the longer we try to cultivate . . . behaviors through measurement and reward, the more damage we do to the quality of . . . relationships, the more we trivialize the meaning of work, and the more disengaged people become."

In chapter 2, Keeping Education in Teaching, Keltchermans centers on the clash between performativity and teacher self-understanding from an international perspective. His policy research undertaken in Belgium includes both European and American examples. It suggests that American teachers are not the only ones struggling with the possibility of the de-professionalization and dehumanization of their practices. Keltchermans tells us that "[teachers] owe it to their students, to the profession, to society and to themselves to resist the pressure of performativity and strive to 'account' for their practice in ways that are educationally relevant (rather than administratively demanded)."

As for Flinders, he cautions us in chapter 3 "that standards cannot take the place of the need for classroom teachers to exercise judgment in the course of their work." He points out that "Today, we make a . . . mistake by assuming that 'higher standards' will automatically increase student achievement when in fact standards alone say nothing about how to meet them and nothing about our capacity to do so." To argue his point, Flinders frames a series of lessons that dismantle the underlying beliefs relating to the need for standards on one hand and the quest for standardization on the other hand.

Readers are invited to carefully ponder these chapter selections. They set the context not only for this book, but the backdrop of teachers' lives, to which we can attest.

CHAPTER 1

The Uses and Abuses of Measurement

Margaret J. Wheatley
The Berkana Institute

Myron Rogers
The Berkana Institute

> Margaret Wheatley, Ph.D., writes, teaches, and speaks about radically new practices and ideas for organizing in chaotic times. She is president emeritus of the Berkana Institute, a charitable global foundation serving life-affirming leaders around the world. Wheatley authored *Finding Our Way: Leadership for an Uncertain Time*, *Leadership and the New Science*, *Turning to One Another: Simple Conversations to Restore Hope to the Future*, and *A Simpler Way* (with Myron Kellner-Rogers).
>
> Myron Rogers, who holds a B.A. in economics, is an author and consultant with a practice in strategic organizational change and leadership development. Over the past 25 years, he has been a consultant to a wide range of public and private sector organizations around the world. He is coauthor of *A Simpler Way* and cofounder of the Berkana Institute.

ABSTRACT

> The work of modern school leaders is to interpret and manipulate numerical views of reality. The problem is that quality teaching and learning are never produced by measurement. They emerge as people voluntarily commit to a shared purpose for educating. Depending on how connected they feel to this mission, teachers and students take responsibility, innovate, and share their learnings with others. We want to use measurement to give education the kind of reality that welcomes its participants to step forward with their desire to contribute, to learn, and to achieve. With this deeper place of understanding, colleagues willingly struggle together in a common work they find meaningful.

This chapter originally appeared in *Finding Our Way* by Margaret Wheatley, published in 2005 by Berrett-Koehler Publishers, Inc.

In the West, we live in a culture that is crazy about numbers. Starting in the sixth century B.C.E., numbers became the means we used to see reality. But over time, numbers became the only reality. Today, we make something real by assigning a number to it. Once it's a number, it's ours to manage and control. The poet W. H. Auden years ago wrote about this Western obsession: "And still they come, new from those nations to which the study of that which can be weighed and measured is a consuming love."

The search for measures has taken over the world as the primary means to control systems and people. We depend on numbers to know how we're doing for virtually everything. We ascertain our health with numbers. How many calories or carbs should I eat? What's my cholesterol reading? We assess people with numbers. What's your IQ? Your EQ? Your GPA? And of course we judge organizational viability with numbers. We manage organizations by metrics to the extent that one executive, when trying to understand a company he had just bought, asked only to see "the pile of metrics you use to run this place."

Numbers are the "hard stuff," the real world of management—graphs, charts, indices, and ratios. And now, increasingly, numbers and standards define and shape the work of educators. Everyone assumes that "you can only manage what you can measure." The work of modern school leaders, who see themselves more as managers than as educators, is to interpret and manipulate these numerical views of reality. Many good principals have been compelled to become earnest students of measurement. But are measures and numbers the right pursuit? Do these measures make for enduring schools? And what effects has this measurement mania created?

This last question is especially important in public education. No Child Left Behind (NCLB) was created from a simple assumption that children would improve their skills if there were a few measures in place, and that teachers could easily improve their teaching with the use of these standards. A second major assumption was that professionals are best motivated by threat and fear, rather than by respect and inclusion. We now have sufficient experience with this legislation to see clearly how these assumptions have played out in practice. As you read through this article, we encourage you to apply the principles outlined here to your experience with the effectiveness of educational standards and especially to NCLB.

Instead of assuming that numbers are the solution, consider this question: What are the problems in our classrooms for which we assume measures are the solution? Presumably, administrators want reliable, quality teaching. They want students to perform better. They want accountability, focus, teamwork, and quality.

If you agree that these are the general ideals you're seeking, ask whether, in your experience, you've been able to find measures that *sustain* these over time.

Do measures actually help your staff turn these ideals into actions? If you haven't found useful measures yet, are you or your staff still expecting to find them, rotating through different reform movements to find the ultimate metric or instrument? Is measurement still believed to be the way to elicit quality performance?

The problem is that the quality of action can never be produced by measurement. Rather, performance capabilities emerge as people feel connected to their work and to each other. They are capacities that emerge as colleagues develop a shared sense of what they want to work on, and as they work in an environment where everyone feels welcome to contribute to that shared purpose.

Each of these dimensions of action—accountability, focus, teamwork, quality—is a choice that people make. Teachers are no different. Depending on how connected they feel to the deeper purpose of the school, they choose to pay attention, to take responsibility, to innovate, to learn and share their learnings. People can't be punished or paid into these behaviors. Either they are contributed or withheld by individuals as *they choose* whether and how they will work with us. Every teacher, in this sense, is a volunteer.

But to look at prevailing educational practices, most leaders consistently choose measurement as the route to action. They agonize to find the right reward to tie to the right measure. How long have schools searched for rewards that will result in better teamwork or more innovation? And haven't leaders noticed that if they find an effective external reward, it only works as an incentive in the short term, if at all? Ironically, the longer we try to cultivate action through measurement and reward, the more damage we do to the quality of our relationships, the more we trivialize the meaning of work, and the more disengaged people become. The commitment to the shared purpose is lost, and this loneliness fails to fire souls toward meaningful action.

Far too many schools have lost the path to quality because they have been burdened with unending measures. Too many teachers have felt compelled to become experts at playing "the numbers game" to satisfy principals and requirements rather than being able to do their jobs. When measurement becomes the primary means of motivating people, it leads us dangerously far from the vital life of schooling that our well-being desperately requires.

But measurement is critical. It can provide something that is essential to sustenance and growth: feedback. All life thrives on feedback and dies without it. We have to know what is going on around us, how our actions impact others, how the environment is changing, how we're changing. If we don't have access to this kind of information, we can't adapt or grow. Without feedback, we shrivel into routines and develop hard shells that keep newness out. We don't survive for long. In any living system, feedback differs from measurement in several significant ways:

Feedback is self-generated. An individual and a system notice whatever they determine is important for them. They ignore everything else.

Feedback depends on context. The critical information is being generated right now. Failing to notice the "now," or staying stuck in past assumptions, is very dangerous.

Feedback changes. What an individual or system chooses to notice will change depending on the past, the present, and the future. Looking for information only within rigid categories leads to blindness, which is very dangerous.

New and surprising information can get in. The boundaries are permeable.

Feedback is life sustaining. It provides essential information about how to maintain one's existence. It also indicates when adaptation and growth are necessary.

Feedback develops fitness. Through the constant exchange of information, the individual and its environment co-evolve toward mutual sustainability.

As we reflect on the capacities that feedback can enhance, it seems we are seeking many similar attributes in our schools. But we haven't replicated the same processes, and therefore we can't achieve the same outcomes. Table 1.1 shows some crucial distinctions between feedback and measurement, as evident in the following contrasts.

If we understand the critical role played by feedback in living systems and contemplate these distinctions, we could develop measurement processes that inspire the action we require, that enhance the vitality and adaptability of education. To create measures that more closely resemble feedback, we suggest the following questions. These are useful design criteria for developing measures and measurement processes.

Who gets to create the measures? Measures are meaningful and important only

Table 1.1 Some important distinctions between feedback and measurement

Feedback	Measurement
Context-dependent.	One size fits all.
Self-determined. The system chooses what to notice.	Imposed. Criteria are established externally.
Information is accepted from anywhere.	Information is put in fixed categories.
The system creates its own meaning.	Meaning is predetermined.
Newness and surprise are essential.	Prediction and routine are valued.
The focus is on adaptability and growth.	The focus is on stability and control.
Meaning evolves.	Meaning remains static.
The system coadapts with its environment.	The system adapts to the measures.

when generated by those doing the work. Any group can benefit from others' experience and from experts, but the final measures need to be their creation. People only support what they create, and those closest to the work know the most about what is significant to measure.

How will we measure our measures? How can we keep measures useful and current? What will indicate that they are now obsolete? How will we keep abreast of changes in context that warrant new measures? Who will look for the unintended consequences that accompany any process and feed that information back to us?

Are we designing measures that are permeable rather than rigid? Are they open enough? Do they invite in newness and surprise? Do they encourage people to look in new places or to see with new eyes?

Will these measures create information that increases our capacity to develop, to grow the shared purpose? Will this particular information help individuals, especially our students and the entire schooling system, grow in the right direction? Will this information help us know whether we are achieving our stated purpose?

What measures will inform us about critical capacities—accountability, learning, teamwork, quality, and innovation? How will we measure these essential capabilities without destroying them through the assessment process?

What processes for measuring serve to develop relationships of trust and support? Most measurement systems drive colleagues apart by applying competitive pressures and individual rewards and punishment. These destroy the very things needed to work well with feedback, such as trust, vulnerability, and openness.

If these questions seem daunting, please be assured they are not difficult to implement. But they do require extraordinary levels of participation—defining and using measures becomes everyone's responsibility. In our experience, we've seen teams, schools, and service organizations where everyone knew that measurement was critical to their success and went at the task of measuring with great enthusiasm and creativity. They were aggressive about seeking information from anywhere that might contribute to those purposes they had defined as most important to their organization, such things as safety, team-based organization, or social responsibility. Their process was creative, experimental, and the measures they developed were often nontraditional. People stretched and struggled to find ways to measure qualitative aspects of work. They developed unique and complex multivariate formulas that would work for a while and then would be replaced by new ones.

They understood that the right measurements gave them access to the information they needed to prosper and grow. But what was "right" kept changing. And in contrast to most organizations, measurement felt alive and vital in these work environments. It wasn't a constraint or deadening weight; rather, it helped people accomplish what they wanted to accomplish. It provided feedback, the information necessary for them to adapt and thrive.

Being in these workplaces, we also learned that measurement needs to serve the deepest purposes of work. It is only when we connect at the level of purpose that we willingly offer ourselves to the organization. When we have connected to the possibilities of what we might create together, then we want to gather information that will help us be better contributors.

But in too many classrooms, just the reverse happens. The measures define what is meaningful rather than letting the greater meaning of the work define the measures. As the focus narrows, teachers and students disconnect from any larger purpose, and only do what is required of them. They become focused on meeting the petty requirements of measurement and, eventually, they die on the job. They end up playing a numbers game and lose motivation to do good work.

If we look closely at the experience of the past few years, it is clear that as a management culture, education has succeeded at developing finer and more sophisticated measures. But has this sophistication at managing by the numbers led to the levels of performance or commitment we've been seeking? And if we have achieved good results in these areas, was it because we discovered the right measures, or was something else going on in the life of our schools?

We would like to dethrone measurement from its godly position, to reveal the false god it has been. We want instead to offer measurement a new job—that of helpful *servant*. We want to use measurement to give us the kind and quality of feedback that supports and welcomes people to step forward with their desire to contribute, to learn, and to achieve. We want measurement to be used from a deeper place of understanding, the understanding that the real capacity of education arises when colleagues willingly struggle together in a common work that they find meaningful.

CHAPTER 2

Teachers' Self-Understanding in Times of Performativity

Geert Kelchtermans
Center for Educational Policy and Innovation

> Geert Kelchtermans, Ph.D., studied educational sciences and philosophy at the Katholieke Universiteit Leuven, in Belgium, where he works as a professor of education and chairs the Center for Educational Policy and Innovation. His research interests are teachers' professional development, educational innovation and school development, micropolitics in teaching, and narrative-biographical approaches and methods in both research and practice of teacher development and teacher education.

ABSTRACT

The dominant discourse in policy, research, and even teaching practice demonstrates an obsession with effectiveness and efficiency as the only relevant criteria. As a consequence, teachers and schools have to account for their performances. This performativity discourse, it is argued in this chapter, however, deeply affects educational practice and one's being a teacher in that it frames the educational relationship as a contractual relationship that is further reduced to instrumental and technical terms. This reductionist approach is criticized by the author, who argues that teacher educators should strive for a different concept of what it means to be a teacher. First, he contends that teachers' conceptions of themselves as teachers—their professional self-understanding—have to be taken seriously and given the central place they deserve in teacher preparation. Further, he argues that education inevitably implies moral choices that can never be fully and indisputably accounted for. Education—and therefore

---

This chapter is based on ideas developed in a keynote lecture presented at the biannual conference of the International Study Association on Teachers and Teaching (Sydney, July 2005). However, the chapter focuses more explicitly on the issues of performativity and accountability and their consequences, and some new practical examples have been added. I would like to thank Betty Achinstein for allowing me to explore the Open Court-Reading Series at the New Teacher Center in Santa Cruz, CA.

being a teacher—implies being vulnerable. Yet as a structural condition of being a teacher, this vulnerability is not only to be endured, but even embraced. In conclusion, the author claims that different forms of accountability are needed in education, forms that do justice to the responsibility and the ethical aspects of the educational relationship, forms that acknowledge that more matters to teaching and being a teacher than issues of effectiveness and efficiency.

Introduction

What kind of teachers do we want to "educate"? What makes a "good teacher"? And who is to define that? What does it mean to be a teacher? How does the policy environment as well as the working conditions in schools determine the teacher education curriculum? And even more importantly, how do teacher educators deal with that environment: Do they take it for granted or critically question the legitimacy of its assumptions?

Teacher educators cannot but ask and answer these questions time and again. When answering them presently, however, they will have to critically take into account the so-called performativity discourse that has become the dominant framework for thinking and talking about education and schooling in policy, in the public debate, as well as in research. This discourse has also resulted in important changes in educational practices. In this chapter, I will briefly discuss this dominant discourse and then focus on a critical analysis of some of its consequences. These consequences are pervasive.

Performativity (and the urge for accountability that goes with it) tend to reduce education and schooling to an almost exclusively instrumentalist and technical agenda, demanding a contractual relationship between teachers (schools) and parents. This not only deeply affects teachers' professional sense of self, but even redefines what counts as "education" (and thus ultimately what is considered to be appropriate teacher education). I will first address the issue of teachers' professional self. Second, I will contend that the educational relationship as an ethical relationship of responsibility cannot be reduced to an instrumental or a contractual one. There cannot be full control over the education process and its outcomes, nor can there be a final justification (accounting) for teachers' actions. Education inevitably contains a dimension of passivity, of things "happening," "taking place," rather than "being made," or "being done." In education there is always something that is happening which is both more and less than intended or planned for. This is experienced by teachers as a sense of vulnerability. Rather than seeing this as a problem, however, I will argue that teachers have to endure this vulnerability and even embrace it for it is exactly this that makes education possible.

Performativity Rules

Over the past two decades, education and schooling have fallen under the spell of performativity, which is:

> a technology, a culture and a mode of regulation that employs judgements, comparisons and displays as means of incentive, control, attrition and change—based on rewards and sanctions (both material and symbolic). The performances (of individual subjects or organizations) serve as measures of productivity or output, or displays of "quality," or "moments" of promotion or inspection. (Ball, 2003, p. 216)

Schools have to perform well, show their effectiveness, and prove that the money invested in them is efficiently used and effectively leads to appropriate output. More and more, education is considered by policymakers, but also by "public opinion," as an investment that has to be efficiently managed and thus has to account for its results. Efficiency and effectiveness have become the ultimate criteria to evaluate education and thus much energy and time have to be spent on monitoring, measuring, and comparing one's students' achievements with that of others (other schools and even other countries).

For this purpose, policy makers—and it must be said, with the eager help of educationalists—have developed a whole series of technologies and procedures: standards and standard-based testing, audit procedures, and methods for self-evaluation, etc. In some countries this has resulted in league tables of schools, high-stakes testing, and scripted curricula. Teacher educators have found themselves confronted with lists of basic competences that they ought to strive for and achieve with their students. But also new teaching methodologies and tools for monitoring, managing, and assessing student teachers' learning are made to fit the performativity frame, different forms of portfolios, assignments for reflection, etc.

Performativity finds its ultimate justification in the quest for educational quality. And this is a very strong justification: Who can be against quality in education? It has become almost impossible to think, talk, and act in education outside the "quality frame." Quality, however, has come to mean the effective and efficient achievement of the standards, and standards claim to be the operational definition of what 'society' wants as a return for its investments in education. The state and the citizens become the critical consumers of the products from the education market, and "if the consumer is supreme, educational values are simply what the consumer happens to want, and it makes no more sense to undertake any great inquiry into those values than into preferences in the matter of cars or brands of chicken tikka" (Blake, Smeyers, Smith & Standish, 2000, p. xii).

Raising standards (as indicators of quality) is the aim and nothing else seems to matter. In this way, the difficult discussion of aims in education is circumvented or postponed. Good is what meets the needs and demands of the client. Underlying this, Ball argues, is "the enterprise form as the master narrative defining and constraining the whole variety of relationships within and between the state, civil society and the economy" (Ball, 2003, p. 226). Not only schools and teachers, but also pupils and parents have come to behave as entrepreneurs, investing time, money, and energy, expecting to receive quality output in return. Education has become a commodity on the increasingly international market (see also Sachs, 2001, p. 155).

Education as a Contractual Relationship

The dominance of this performativity in education, however, has a profound impact as it radically repositions teachers and schools on the one hand, and pupils and parents on the other. Both "parties" nowadays find themselves positioned in an economical, contractual relationship (Masschelein & Simons, 2002). A contractual relationship means that both parties are linked through an explicit set of rights and duties. It furthermore means that teachers and schools become producers who can be held accountable by the consumers (the pupils and parents, or—on their behalf—the state). It is assumed that educational relations can be made transparent, that outcomes can be explicitly identified, attributed, measured and compared, and that education and teaching become acts of "producing," or more specifically technical and instrumental acts of connecting the most efficient means to ends.

The impact of this way of thinking on what it means to be a teacher, on what counts as "education," on what is relevant to be studied by researchers, on teachers' professional relationships, etc. is pervasive (see Achinstein & Ogawa, 2006; Jeffrey, 2002; Troman, 2000; Woods & Jeffrey, 2002). As a consequence of this apparent contractual relationship, enterprising ways of conceiving education have become "self-evident." The enterprise as "master narrative" (Ball, 2003) prevails.

As a matter of fact, this idea of a contract is to be understood not only metaphorically, but also in the literal sense of the word as one sees illustrated in different forms of "learning agreements" (for example in higher education) between students and the institute or school (Masschelein & Simons, 2002, p. 569):

> Learning agreements of this kind are typically required to contain (or pretend to contain):
>
> - a precise and transparent formulation of the conditions and requirements that have to be fulfilled by the students (at the beginning and the start of the learning process-start and end competences)
> - a description of the instruments (for example, the assessment tools) through which the student can evaluate whether or not she meets these requirements
> - an indication of the kind of support that is offered by the institution and the teachers, and to which the student has a right

At the micro-level, one can also see it illustrated in scripted curricula, like Open Court, and the ways they are implemented with their emphasis on "fidelity," pacing schedules, and accountability sanctions. The curriculum consists of an expensive set of materials that provides a very detailed prescription of the different activities to be performed by the teachers (and students): 10 teacher guides per teacher, reading books, instructions about how to arrange a classroom or what materials to hang on the walls, additional resource information, OHP-slides, worksheets (with the correct answers in the teachers' manual, even providing essential information like "students' answers may vary" or "students may use a second sheet of paper if insufficient space," which are written in red, like a warning to the teacher about student variation from agreed upon scripts), assessment tools and tests, very specific planning schedules, and so on.

Although one could rightly argue that this material is a valuable aid for teachers to deal with the complexities of the classroom, one should at the same time ask what concept of "teacher professionalism" is communicated here by the policies of fidelity that monitor and enforce strict adherence to, and no questioning of, prescribed texts (see Achinstein & Ogawa, 2006). Teachers following this scripted curriculum hardly have to do any thinking themselves—not even in terms of adapting the materials or the learning activities to the particular characteristics and needs of their actual students and the particular circumstances in the school. In some districts they are not even allowed to make any decisions that differ from the curriculum prescriptions. In this way, teaching and education are being reduced to the technically correct and rigorous execution of prescriptions. It echoes Apple and Junck's (1992) article title of fifteen years ago, "you don't have to be a teacher to teach this unit."

The consequences of this approach are even more pervasive in a policy environment where high-stakes testing is directly linked to the implementation (= the execution) of the curriculum. The straightforward rationale sounds like this: Successful teaching = execution of the curriculum prescriptions = reflected in students' test scores. This policy rationale is justified by referring to the contractual duties the teachers and schools have toward their clients (parents and students, or—on their behalf—the district or the state), who from there derive the right to evaluate the outcome of their actions and possibly sanction them. Achinstein and Ogawa (2006), for example, documented the shocking experiences of beginning teachers who—out of educational commitment to meet their students' needs—deviated from the scripted curriculum and eventually were heavily sanctioned (and lost or left their jobs).

Both as a researcher and as a teacher educator, I find these developments deeply disturbing since they radically alter teaching and education. Or, put more specifically, framing the educational relationship in contractual (economical) and technical terms not only reduces, but even changes what counts as "education" and therefore changes what it means to be a teacher. It results in being blind to, ignoring, or even denying important aspects of the educational reality. They are considered to be not relevant, not valuable, and therefore "ignorable." My suspicion, however, is that in this way, not only are essential aspects of teaching and education being lost and moved out of sight, but by doing so we neglect—under the banner of a quest for quality—those aspects that constitute powerful sources of motivation, commitment, and job satisfaction for teachers. As such, our attempts to improve quality may paradoxically result in undermining exactly that for which we strive.

We are now facing the question: Where are we in all this as teacher educators or educational researchers? The technologies that make up the actual reality of performativity could not have appeared without numerous educationalists, researchers, and teachers devoting themselves to the performativity agenda, which is quite understandable. The idea of being able to design and define the exact educational processes leading to effective outcomes is very appealing. It even entails the promise of undoubtedly proving the scientific value of the discipline, the professional expertise of educationalists, and so forth. In other words, there is a strong appeal stemming from what Sachs (2001) has called "managerial professionalism."

Yet, if we do not want to simply be accomplices, we have to critically ask whether and in what respect our activities as teacher educators and researchers contribute to the negative impact of the performativity agenda. This certainly is a political issue, but it is also a fundamental educational issue about responsibility, commitment, and professionalism.

Reflections from a Narrative-Biographical Frame

My reflections on these issues are guided by my narrative-biographical research on teachers' professional development (Kelchtermans, 1993, 1996, 2005; Kelchtermans & Hamilton, 2004), as well as the use of (auto)biographical reflection in teacher education and in-service training. Collecting and carefully analyzing the narrative accounts in which teachers make sense of their career experiences has proven to be a powerful approach to disentangling and understanding the complexities of teachers' lives and work.

The narrative-biographical approach acknowledges the dynamic, interactionist, and contextualized character of education. As such, it challenges the dominant discourse of performativity that reduces education—and the people involved in it—to an instrumental enterprise. The challenge we are facing as teacher educators and researchers is to develop conceptions of teaching and being a teacher that go beyond the contractual relationships and acknowledge the central place of commitment as a person at the heart of teacher professionalism. Before I argue that stance more systematically, a few examples from both my experiences as a teacher educator and as a researcher may illustrate my point.

OBSERVATION 1: WHAT MAKES A GOOD/BAD TEACHER?

At the end of the teacher education program at the university, just after the student teachers finish their practical training (internship), I offer a seminar on autobiographical reflection. One exercise in this seminar is the assignment to recall their best and worst teacher ever. I ask the students to depict the person in a short narrative vignette and provide a title for the piece. Next, I ask them to analyze that experience and make explicit what it was that made this particular teacher so significant to them (either positively or negatively). To the students' own surprise, they often find themselves to have very vivid and detailed memories.

Time and again the stories in the vignette reveal people who appeared in the students' lives as a person. Positive examples mention the passionate subject teacher who was so deeply driven by interest and knowledge of the subject that she managed to trigger the students' curiosity, get them really interested, and motivated them to work and study for a subject they normally did not like at all. Or the teacher who subtly showed attention and interest in the individual student, and made her believe in herself and value herself at a time when she was struggling with questions of identity as an adolescent. Quite often those teachers were pedagogically not exactly "examples of good practice"—i.e. had messy schemes on the blackboard; lack of structure in their course materials and teaching, etc. Although the student teachers did acknowledge the value and importance of those technical skills, their former teachers were "excused" for not enacting them because their other characteristics were valued far more.

The negative cases in students' stories exemplify the opposite. A common element in the stories, and a comment that I also have often heard from experienced teachers, is "I want to make a difference in the students' lives as a person." This "difference" refers to issues that go well beyond curriculum prescriptions, output testing, etc. It seems that not only as Russell (1997) argues, "how I teach is the message," but we have to go even further and acknowledge that "who I am as a teacher is the message."

OBSERVATION 2: CAN TEACHERS KNOW THEY ARE MAKING A DIFFERENCE?

In my narrative-biographical work with experienced primary school teachers, one of the issues that they almost all struggled with (or had struggled with) was finding a balance between internal and external attribution of student outcomes. Since student outcomes—especially in the climate of increased pressures for accountability—are being used as the indicator for the quality of the teachers' work, the way they attribute the causes for those outcomes is highly relevant to their job motivation. All teachers realize that student outcomes are only partially determined by their teaching. Equally or sometimes more decisive are personal factors (motivation, perseverance, etc.) or social factors that are hard to influence, change, or control. These factors create ambivalence among the teachers.

Teachers with a high internal "locus of control" may experience high job satisfaction when student outcomes are good. On the other hand, when pupils' learning outcomes are poor, they may tend to blame themselves and feel frustrated and ineffective. Teachers with a high external "locus of control" often ascribe student outcomes to factors beyond their efforts and often beyond their control. This may then have a negative impact on their personal feelings of professional competence (I can't make a difference, so why bother . . . ?) and thus negatively affect their motivation, and eventually their self-esteem. During their careers, teachers find themselves challenged to properly balance between the internal and external locus of control, between a satisfying sense of efficacy and a realistic acknowledgement of one's limited impact, and between exhausting personal commitment and cynical disengagement (see also Huberman, 1989).

Understanding Teachers' Self-Understanding

The central role of teachers' sense of self in understanding their professional actions has been acknowledged for a long time and by many authors:

> the teacher as a person is held by many within the profession and outside it to be at the centre of not only the classroom but also the educational process. By implication, therefore, it matters to teachers themselves, as well as to their pupils, who and what they are. Their self-image is more important to them as practitioners than is the case in occupations where the person can easily be separated from the craft. (Nias, 1989, pp. 202–203)

Studying teachers' professional lives, Ball and Goodson (1985) have argued that "the ways in which teachers achieve, maintain, and develop their identity, their sense of self, in and through a career, are of vital significance in understanding the actions and commitments of teachers in their work" (p. 18). The idea that it matters in teaching who the people are is also illustrated in the observations cited. Teachers' concerns with making a difference in students' lives, both as a person and as a "teacher," explain why they struggle with the attribution of student outcomes (locus of control). If they were not involved as a person, they just would not care!

Components of Teachers' Self-Understanding

In the career stories I collected from experienced teachers (Kelchtermans, 1993, 1996) the issue of the "self" or the teacher's sense of "identity" is prominent. Nias (1989) was right when she observed teachers' "persistent self-referentialism," the fact that when talking about their professional actions and activities, teachers cannot help speaking about themselves (p. 5). The analysis of this "self-referentialism" in teachers' narrative accounts of career experiences brought me to a more differentiated concept of that "self" or that sense of "identity." I have, however, purposefully avoided the notion of "identity" because of its association with a static essence, implicitly ignoring or denying its dynamic and biographical nature (development over time). Instead I have used the word "self-understanding." The term refers to both the understanding one has of one's "self" at a certain moment in time (the result), as to the fact that this result is always preliminary as it stems from an ongoing process of making sense of one's experiences and their impact on the "self."

In my analysis of teachers' career stories, I identified five components that make up teachers' self-understanding: self image, self-esteem, job motivation, task-perception, and future perspective (Kelchtermans, 1993, 1996). Self-image is the descriptive component, the way teachers typify themselves as teachers. The evaluative component, or the self-esteem, refers to the teacher's appreciation of his/her actual job performances ("how well am I doing in my job as a teacher?").

Both the example of teachers' struggles with internal/external attribution of students' achievements and the one on "best/worst teacher" illustrate this evaluative component. Job motivation (or the conative component) refers to the motives or drives that make people choose to become a teacher, to stay in teaching, or to give it up for another career. The normative component of task perception encompasses the teacher's idea of what constitutes his/her professional program and his/her tasks and duties in order to do a good job. It reflects a teacher's personal answer to the questions: What must I do to be a proper teacher?; What are the essential tasks I have to perform in order to do well?; What do I consider as legitimate duties to perform?; and What do I refuse to accept as part of "my job"?

This normative component was illustrated by a spontaneous but intense discussion I witnessed last year in one of the autobiographical seminars, which developed around the question: "Should I give students my mobile phone number or not?" Whether or not this was formally required by their job description or by the school's internal rules was never a part of the conversation, nor was this a matter of competences and skills. What was at stake in this question was the issue of "availability": To what extent do I feel I have to be available for my students? And why? How can I balance committed work as a teacher, with still being able to have a private life of my own?

This experience reveals a question that deeply engages student-teachers (and as such proves to be very important to them) that has nothing to do with efficiency, standards, or effective teaching. It rather refers to a quality of the interpersonal relationship, as well as to the kind of person they want to be as teachers. Although from a distance it may seem very trivial, it nonetheless shows which fundamental value-choices are at stake: To what extent do I feel I need to be available for my students? It illustrates at the same time that even scripted curricula, formal job descriptions, or school regulations can never take away the need for professional teachers to make this kind of moral judgements and commitments.

Finally, self-understanding also includes a time-element: The future perspective reveals a teacher's expectations about his/her future in the job ("how do I see myself as a teacher in the years to come and how do I feel about it?"). This is implicitly illustrated in the example with the mobile phone: If I want to be able to continue in the job and to be healthy, I will have to find a work-life balance. While these five components can be distinguished independently, all are intertwined and refer to each other.

The Critical and Political in Teachers' Self-Understanding

Because in self-understanding the issue at stake is not a neutral statement but one's self and the moral choices and emotions it encompasses, the narrative accounts always entail an aspect of negotiation (seeking recognition or acknowledgement). The value-laden choices in the task perception, for example, can be contested, and engaging in a discussion about them thus entails the risk of finding one's deeply held beliefs and assumptions challenged or questioned. Yet, at the same time, these discussions are very powerful opportunities for (student) teachers to become explicit about themselves and their professional self-understanding, to reflect, articulate, and communicate on them, and to learn from other (student) teachers' self-understandings about their perspectives and fundamental choices.

I agree with Sachs (2001) who claims that "making these narratives public is a source for lively professional development. . . . Critical self-narratives about professional identity at the individual and collective level have clear emancipatory objectives" (p. 158). Explicitly discussing one's self-understanding is the only way that teachers in a school can negotiate and develop shared understandings and shared moral and political choices as colleagues (and this of course applies to teacher educators as well). In teacher education programs, this topic is a very powerful one to reflectively deepen one's learning experiences during and after internship.

It is obvious, however, that this idea of self-understanding and its link with professional development stands in great tension with the logic of performativity. For example, scripted curricula strongly convey the message that a teacher as a person doesn't matter (since anyone could teach the materials). It states that not only are teachers interchangeable, they ultimately are not even allowed to "matter" in order to make sure that all students get a quality education and "no child is left behind." In this way, teachers are not supposed to have a voice in defining good education, nor in deciding what is in the best interest of particular students at a particular time.

The narrative-biographical perspective argues for the opposite view. Teachers as persons do matter, as does their agency and responsibility to make choices (in the context) and to act. Teachers also differ as do the contexts in which they work. Ignoring or denying these differences just means downplaying crucial aspects of what constitutes education. So, when preparing future teachers, the idea that one's self-understanding as a teacher is an inherent element in teachers' professionalism ought to get extensive and explicit attention in the curriculum (for example in assignments for reflection on practical training experiences).

The Pedagogical Quality of Vulnerability

Teachers' self-understanding develops in interaction with the context in time and space. Yet the most immediate context for teaching is the relationship between teachers and their students. This constitutes the very heart of teaching and thus of a "good education."

THINKING BEYOND THE CONTRACTUAL AND THE TECHNICAL

The examples given point to dimensions in the educational relationship that cannot be properly typified as a contract or as technical or instrumental issues. The discussion concerning the mobile phone reveals the issue of having to decide about one's availability as a teacher. This discussion is not technical, instrumental, or functional, but it refers to the felt need by (student) teachers to be present and available for their students in a particular way. One feels one owes this to the pupils, that they are entitled to claim it, and yet the student-teachers are well aware that there have to be limits to one's availability if one wants to survive in the job and remain healthy.

Also, balancing the internal and external attribution of student outcomes echoes the same dimension; there is no clear answer and the balance has to be re-established time and again. But, that balance doesn't leave the teacher indifferent, since intense emotions, such as powerlessness, self-doubt, and frustration, go with it. The feeling of being unable to make a difference in pupils' lives is devastating for teachers' job motivation, but also—and equally important—for their job satisfaction and commitment.

What we tend to overlook in our eagerness to influence, to improve and to show our instrumental effectiveness as teachers and teacher educators, is that the educational relationship is not merely constructed or intentionally built. There is a dimension to it that escapes our control, our intentions, our planning. In other words, the educational relationship is a being-together-with-others in which one finds oneself (Masschelein & Simons, 2002). It is a situation of "being exposed" to the other, a dimension of passivity. This phrasing is important, since in English the expression "finding one's-self in" nicely captures the fact that it is not something that is being done to the teachers, but rather a fundamental characteristic of education.

I have referred to this dimension in teachers' job experience as "vulnerability" (Kelchtermans, 1996, 2005), concluding from my research that "the basic structure in vulnerability is always one of feeling that one's professional identity and moral integrity, as part of being 'a proper teacher,' are questioned and that valued workplace conditions are thereby threatened or lost. Coping with this vulnerability therefore implies political actions, aimed at (re)gaining the social recognition of one's professional self and restoring the necessary workplace conditions for good job performance" (Kelchtermans, 1996, p. 319).

The experience of vulnerability results from the fact that teachers do not feel in control of what they considered to be valued working conditions (infrastructure, contract, professional relationships). Policy measures and imposed educational reforms that are not congruent with their deeply held beliefs about good teaching, but from which teachers felt they could not escape, further contribute to the experience of vulnerability (see also Nias, 1999, p. 226; Van den Berg, 2002, p. 577).

VULNERABILITY AS A STRUCTURAL CONDITION IN TEACHERS' WORK

Vulnerability, I argue, should not primarily be understood as an experiential category, but rather as a structural condition teachers (or educators in general) find in themselves. Teaching implies an ethical relationship of responsibility in which one engages oneself as a person. This commitment cannot be properly conceptualized as a merely contractual or intentional relationship (see also Ball, 2003; Jeffrey, 2002). There is more to teaching and being a teacher than technically linking the means (teaching actions and methods) that promise to be most effective to the ends. Although this instrumental concern in the teachers' job is a legitimate dimension, there is always more at stake.

Since the relationship with students is an ethical one (Fenstermacher, 1990, p. 132), the teacher never has full control over the situation, nor over the outcomes of his/her actions. In spite of thoughtful planning and purposeful skilled action (however important they are), the pedagogical relationship can never be fully controlled, nor can one be sure that one's actions will convey the meaning they were intended to have for the students. As such, the educational relationship implies a dimension that radically escapes control and intervention; therefore, the relationship contradicts the fundamental activist bias in educational theory and intentional actions of teaching because of its taken for granted association of doing something and bringing something about.

This "entrepreneurial" (Masschelein & Simons, 2002) interventionist root metaphor is so strong that it makes it almost impossible to see and acknowledge the aspects of "passivity," of "being exposed" to the other, of "finding oneself in a situation" in which things "can happen, can take place" (instead of "being done"), that are also intrinsically part of educational relationships. The entrepreneurial metaphor operates as a powerful, discursive pattern constituting the horizon of how one can think or speak about education.

Since a teacher or educator, because of the fundamental ethical character of the relationship, can never fully prove the effectiveness of his/her actions and since there is no uncontested moral stronghold to justify one's specific actions, being a teacher implies that one's actions and decisions can always be disputed. In other words, this vulnerability is a structural condition that constitutes the specific character of the educational relationship and therefore also of teachers' professional self-understanding. Teachers, however, as Van Manen (2002) argues, tend to focus on the pedagogical, the complexity of relational, personal, moral, and emotional aspects of their everyday acting with children or young people that they teach. He concludes that "pedagogy is the condition for the instructional dimension of teaching . . . pedagogy makes the practice of teaching possible in the first place" (Van Manen, 2002, p. 137).

The lack of full control, the fact that accountability procedures either neglect or make instrumental (and thus reduce) the interpersonal dimension in teaching, and the absence of an ultimate ground for justifying one's actions as a teacher—I would argue—are realities that teachers have to endure: "To teach is to be vulnerable . . . to be vulnerable is to be capable of being hurt" (Bullough, 2005, p. 23). And this is not a comfortable condition to live in. It also explains why some teachers hold a rather positive stance toward standards and standardized testing. Standards and tests promise certainty or a final proof of one's "quality" as a teacher—even if it is a delusory certainty that demands a very reductionist understanding (and experiencing) of the educational relationship and one's role in it.

On the other hand, the condition of vulnerability is at the same time that which constitutes the very possibility for the "pedagogical" (see Van Manen, 2002) to happen in the interpersonal relationship between teachers and pupils. The relationship of an ethical and thus vulnerable commitment opens up the chance that education (literally) "takes place." This is what clearly shines through in (student) teachers' narratives about what made their "best teachers" so meaningful. These encounters between student and teacher make the teacher feel that s/he is really "making a difference as a person" in the student's life. Joy, pride, and existential personal fulfilment are the emotions that go with it. And, it goes without saying, that these experiences boost job motivation and commitment and therefore contribute to "good education."

For these reasons, vulnerability is not only a condition to be endured, but one that is also to be acknowledged, to be cherished, and even embraced by teachers. Policy measures and procedures that ignore or downplay this aspect of vulnerability in teachers' work lives may therefore take away or threaten exactly those experiences on which not only teachers' job satisfaction and commitment rest, but that also ultimately are the conditions that strongly contribute to student outcomes. Therefore, teacher educators and educational researchers ought to persist in carefully and critically questioning accountability/performativity policies on their rationale and effects.

The Critical Promise of the Narrative

Both practical experiences and research have shown the strong potential of narratives and (auto)biographical reflection, not only in research on teaching, but also in teacher education (Carter & Doyle, 1996; Casey, 1995–1996; Clandinin, 2006). Narrative language is "the discourse structure in which human action receives its form and through which it is meaningful" (Polkinghorne, 1988, p. 135). Maybe the most fundamental contribution of narrative and biographical approaches to teacher education and educational research is that they bring with them a different language that allows for the non-technical dimensions of teaching and being a teacher to be conceptualized, talked about, and critically challenged.

Sockett and LePage (2002), for example, argue for the urgent need to develop a language to address the moral dimensions of teaching: "Teachers do not lack moral sophistication because they are not moral people. Just the opposite, most teachers are drawn to teaching because of their moral commitments. Moral language is missing in classrooms: but it is also missing in the seminar rooms and lecture halls of teacher education" (Sockett & LePage, 2002, pp. 170–171).

The moral language may be missing because the discourse has to come from teacher educators whose discourse about teaching needs to include reference to the moral and a consideration of what is moral, which can be painful. For some, it is far more embarrassing to be accused of not walking their talk about moral action than their talk about technical, skill-related action. For example, a teacher educator might rather have a student say "you are lecturing us about constructing our own knowledge" than hear how an interaction lacks integrity. Continuing to talk at the level of technical competence rather than the moral makes the person less vulnerable (Kelchtermans & Hamilton, 2004, p. 795). In narrative-biographical language, moral dilemmas, emotional experiences, and political struggles can find a place in it and can thus be acknowledged as fundamental to the experience of teaching and being a professional teacher.

Conclusion

These reflections on the dominant performativity/accountability discourse and its consequences for what counts as education, or for what it means to be a teacher, show the need for anyone involved in teacher education, induction, or in-service training to strive for a concept of teaching and being a teacher that does justice to the complex reality of education. Although the importance of thorough knowledge (on subject matter, educational theory, teaching methods) and skilful action for professional teaching are indisputable (and should therefore be carefully dealt with in teacher education), they are not sufficient. The preparation of teachers should also include thoughtful reflection on one's developing a professional self-understanding, as well as the acknowledgement of one's vulnerability and the fact that education per se entails a dimension of passivity, of things "happening" rather than "being done." In this way, teacher education does justice to the ethical appeal for responsibility and commitment that is central in "education." That way teacher education really prepares prospective teachers for a "qualified" and "realistic" entry into the profession.

Professional teachers master the skills to critically question working conditions and their own actions in it, as well as to challenge the taken-for-granted assumptions in policy and administrative regulation. Elsewhere (Kelchtermans & Ballet, 2002a, 2002b), I have argued for the need to develop micro-political literacy as part of teacher professionalism: the capacity to "read" professional situations in terms of different interests, the skill to navigate and negotiate these working conditions, and the commitment to act out of a responsibility for what is educationally in the best interest of the student, even if the student teacher remains vulnerable to questioning or debate.

Policymakers and administrators should take up the challenge to conceive of curricula, teacher education programs, and so forth, but also should dwell on structural working conditions for teachers that acknowledge the inevitable complexity of education. They should move beyond the illusion that formal and standardized prescriptive and repressive procedures would contribute to good education.

Teacher educators and researchers should have the courage to persevere in their critical quest for non-simplistic answers to the difficult eternal questions in education and not yield to the tempting illusions nor to the threats of a reductionist managerial and instrumentalist discourse from those who are formally in power. They owe it to their students, to the profession, to society, and to themselves to resist the pressure of performativity and strive to "account" for their practice in ways that are educationally relevant (rather than administratively demanded).

Those "accounts" will have to entail teachers' thoughtful choices and commitments in full awareness of the vulnerability that characterizes them. Those "accounts" cannot be reduced to evidence of effectiveness as formally defined and required by others (for example, through high-stakes testing). They will have to reflect the responsibility that is taken up: the thoughtful and committed "responses" teachers give to the educational demands placed on them by their students. This is not a romantic, naïve, or idealistic claim, but rather a very realistic one, one that does not give in to inappropriate reductionist conceptions of teaching and learning, but rather takes the "reality" of education seriously and looks for the proper language to speak out on the issues.

References

Achinstein, B., & Ogawa, R. (2006). (In)Fidelity: What the resistance of new teachers reveals about professional principles and prescriptive educational policies. *Harvard Educational Review*, 26(1), 30–63.

Apple, M. W., & Jungck, S. (1992, 1996). You don't have to be a teacher to teach this unit: Teaching, technology and control in the classroom. In A. Hargreaves & M. G. Fullan (Eds.), *Understanding teacher development* (Teacher Development series) (pp. 20–42). London: Cassell.

Ball, S. J. (2003). The teacher's soul and the terrors of performativity. *Journal of Education Policy*, 18(2), 215–28.

Ball, S., & Goodson, I. (Eds.) (1985). *Teachers' lives and careers*. London-Philadelphia: Falmer Press.

Blake, N., Smeyers, P., Smith, R., & Standish, P. (2000). *Education in an age of nihilism*. London-New York: Routledge/Falmer.

Bullough, R. V., Jr. (2005). Teacher vulnerability and teachability: A case study of a mentor and two interns. *Teacher Education Quarterly*, 32(2), 23–40.

Carter, K., & Doyle, W. (1996). Personal narrative and life history in learning to teach. In J. Sikula, T. J. Buttery, & E. Guyton (Eds.), *Handbook of research on teacher education*. Second edition (pp. 120–42). New York: Macmillan.

Casey, K. (1995, 1996). The new narrative research in education. *Review of Research in Education*, 21, 211–53.

Clandinin, J. (Ed.). (2006). *Handbook of narrative research methodologies*. Thousand Oaks: Sage.

Fenstermacher, G. (1990). Some moral considerations on teaching as a profession. In J. Goodlad, R. Soder, & K. Sirotnik (Eds.), *The moral dimensions of teaching* (pp. 130–51). San Francisco: Jossey Bass.

Huberman, M. (1989). The professional life cycle of teachers. *Teachers College Record*, 91(1), 31–57.

Jeffrey, B. (2002). Performativity and primary teacher relations. *Journal of Education Policy*, 17(5), 431–546.

Kelchtermans, G. (1993). Getting the story, Understanding the lives. From career stories to teachers' professional development. *Teaching and Teacher Education*, 9(5/6), 443–56.

Kelchtermans, G. (1996). Teacher vulnerability. Understanding its moral and political roots. *Cambridge Journal of Education*, 26(3), 307–23.

Kelchtermans, G. (2005). Teachers' emotions in educational reforms: Self-understanding, vulnerable commitment and micropolitical literacy. *Teaching and Teacher Education*, 21, 995–1006.

Kelchtermans, G., & Ballet, K. (2002a). The micro-politics of teacher induction. A narrative-biographical study on teacher socialisation. *Teaching and Teacher Education*, 18(1), 105–20.

Kelchtermans, G., & Ballet, K. (2002b). Miropolitical literacy: Reconstructing a neglected dimension in teacher development. *International Journal of Educational Research*, 37, 755–67.

Kelchtermans, G., & Hamilton, M. L. (2004). The dialectics of passion and theory: Exploring the relation between self-study and emotion. In J. Loughran, M. L. Hamilton, V. Kubler LaBoskey, & T. Russell (Eds.), *The international handbook of self-study of teaching and teacher education practices* (pp. 785–810). Dordrecht: Kluwer Academic Publishers.

Masschelein, J., & Simons, M. (2002). An adequate education for the globalized world? A note on the immunization of being-together. *Journal of Philosophy of Education*, 36(4), 565–84.

Nias, J. (1989). *Primary teachers talking. A study of teaching as work*. London: Routledge.

Nias, J. (1999). Teachers' moral purpose: stress, vulnerability and strength. In R. Vandenberghe & M. Huberman (Eds.), *Understanding and preventing teacher burnout. A sourcebook of international research and practice* (pp. 223–37). Cambridge, UK: Cambridge University Press.

Polkinghorne, D. (1988). *Narrative knowing and the human sciences*. Albany: State University of New York Press.

Russell, T. (1997). Teaching teachers: How I teach is the message. In J. Loughran & T. Russell (Eds.), *Teaching about teachers: Purpose, passion and pedagogy in teacher education* (pp. 32–47). New York: Falmer Press.

Sachs, J. (2001). Teacher professional identity: competing discourses, competing outcomes. *Journal of Educational Policy*, 16(2), 149–61.

Sockett, H., & LePage, P. (2002). The missing language of the classroom. *Teaching and Teacher Education*, 18(2), 159–71.

Troman, G. (2000). Teacher stress in the low-trust society. *British Journal of Sociology of Education*, 21(3), 331–53.

Van den Berg, R. (2002). Teachers' meanings regarding educational practice. *Review of Educational Research*, 72(4), 577–625.

Van Manen, M. (2002). Introduction: The pedagogical task of teaching. *Teaching and Teacher Education*, 18(2), 135–38 (special issue: "The pedagogical task of teaching").

Woods, P., & Jeffrey, B. (2002). The reconstruction of primary teachers' identity. *British Journal of Education*, 23(1), 89–106.

CHAPTER 3

Standards and Accountability
WHAT SHOULD TEACHERS KNOW?

David J. Flinders
Indiana University at Bloomington

> David J. Flinders, Ph.D., is an associate professor in the Department of Curriculum and Instruction at Indiana University, Bloomington. His research interests focus on school reform and the professional lives of classroom teachers. He is the co-editor (with Stephen J. Thornton) of *The Curriculum Studies Reader*, now in its second edition.

ABSTRACT

> This chapter identifies four lessons from research on the nature and impact of standards-based accountability. The first lesson is that standards do not *cause* learning. The second lesson is that standards cannot replace the need for classroom teachers to exercise judgment in the course of their work. The third lesson is that accountability alone will not solve the achievement gap between white and minority students. On the contrary, today's accountability policies may actually widen this gap. The fourth lesson is that teachers are accountable first to their students. All four lessons further an understanding of the limitations of accountability and thus the need for alternative reform efforts. Moreover, the final lesson is especially important as a basis for re-thinking accountability in ways that will allow teachers to enhance their contributions to student learning.

How should we prepare teachers in an era of standards-based accountability? What do they need to know to think critically about standards, to discuss them intelligently, and to use them wisely? This chapter identifies four lessons that stem from research on the accountability reforms of the past decade. In the U.S., the first two lessons are aimed at a better understanding of what standards are. In today's information age, the standards of a particular content area and grade level are usually available to teachers online. In this sense, it is easy to say that teachers should know the standards relevant to their work. But what exactly are standards? How are they distinguished from goals or values? What does it mean to say that a standard has been met? Moreover, what is the relationship between standards and student learning? The third lesson discussed in this chapter is aimed at understanding another but equally important relationship—that is, the relationship between accountability and educational equity. In particular, what are the effects of accountability on the achievement gap between white and minority students? The final lesson is intended to help teachers rethink notions of accountability in ways that enhance their relationships with students and their contributions to student learning.

These four lessons are developed from the underlying thesis that education has little to gain if we respond to accountability systems only by compliance and capitulation. Some educators already seem to have reached a point of utter resignation. While undoubtedly a source of frustration, the accountability movement is, nevertheless, also an opportunity to think critically and constructively about concerns that are central to classroom teaching. We teacher educators, for example, hope that the future teachers with whom we work will someday model for their own students certain dispositions including independence of thought, respect for systematic inquiry, open-mindedness, and the habit of careful reflection. Our own responses to standards and accountability should be occasions for strengthening these dispositions.

Lesson #1: Standards do not cause student learning. The notion that standards can be used to strengthen the academic rigor and learning outcomes of educational programs pre-dates the No Child Left Behind (NCLB) Act by more than a decade. Goals 2000 and earlier federal policies reflected the standards and accountability movements as they gained strength in the 1990s. The taproot of this movement leads directly back to the highly critical report of education, *A Nation at Risk* (National Commission on Excellence in Education, 1983; see also Noddings, 2003, chapters 4 and 10). Yet, close parallels to today's trends can be found even earlier. What pass for standards today are very similar to the "educational objectives" of yesterday. In the 1960s and 1970s, writers such as Mager (1962) and Popham (1970) argued that if teachers would unambiguously state their lesson objectives upfront, before instruction, students would know exactly what they were supposed to learn, and this knowledge would improve learning outcomes. Today's standards are believed to play a similar role by specifying the content and skills that students are intended to learn.

The objectives movement of the 1960s and 1970s may well have helped some teachers think more clearly about their intended outcomes. However, this much-heralded technique did not turn out to be the panacea that would-be reformers had hoped. While the basic logic seemed a matter of common sense (that we should decide our destination before setting out on the journey), teachers were not enthusiastic about spending hours each week writing out detailed lesson objectives. And once written (or obtained from an "objectives bank"), having definite aims at hand did not seem as useful as expected. Most teachers already knew where their teaching was headed, and stating objectives in minute detail did not get them any closer. Today we make a similar mistake by assuming that "higher standards" will automatically increase student achievement when in fact standards alone say nothing about how to meet them and nothing about our capacity to do so.

Another similarity between the objectives movement and contemporary accountability trends is that objectives seemed more readily applied to some types of learning rather than others. Educators could easily specify objectives when it came to factual information, rote memorization, and simple skills learning. Yet, in areas such as the arts, music, literature, historical interpretation, and critical thinking, educational objectives are more difficult to pin down (Eisner, 1967). This unevenness of applicability has again surfaced in standards-based accountability systems, particularly with respect to the arts and elective subjects at the secondary level.

Leslie Santee Siskin (2003a), for example, examined the impact of state accountability measures on subjects outside of the "core" requirements for high school graduation in four states. Standards, Siskin notes, legitimize a subject area as important to the school. Nevertheless, this legitimization often turns out to be a backhanded compliment when subjects are ranked relative to one another on the basis of their worth. Siskin reports that in Kentucky, for example, music is included as a tested subject within the humanities. Overall, the humanities count for 7.13 percent of a school's total score, whereas English alone counts for 14.25 percent (p. 88). How are students to interpret a subject's relative worth carried out to the second decimal point? Siskin also reports that music teachers have had to adjust their teaching and curriculum in ways that accommodate a paper-and-pencil test, thereby lowering their standards rather than raising them. Siskin concludes by questioning whether or not high standards are even possible, let alone practical, across all subject areas.

For many subjects, standards-based accountability is a lose-lose proposition—damned if you are tested and damned if you are not. But even if the arts, foreign languages, and other areas suffer as a result, won't these sacrifices be offset by gains in student achievement overall? Or, to state this question another way, aren't standards likely to fare better than did educational objectives with respect to their impact on classroom teaching? I have already expressed skepticism on this point. However, belief that standards make a difference is fueled by an increasingly widespread misconception that prior to the most recent wave of accountability measures, schools were not held accountable for the education of their students. According to this scenario, supposedly unaccountable schools could be threatened or shamed into reform.

Both assumptions are questionable. First, the vast majority of American schools have long been externally accountable to school boards, parent associations, and accreditation agencies. Most schools are also "internally" accountable through professional development and teacher licensing policies. Second, holding a gun to a person's head does not make him or her a better person. In education, researchers tell us that the causes of learning rest with a host of complex factors that include school resources, teacher retention, community support, parent involvement, class size, student health and nutrition, prenatal care, and student mobility (Barton, 2004a). Standards-based accountability has such a small influence on any of these factors that it would indeed be surprising if student achievement were to directly benefit from today's accountability measures.

This prediction is supported by the early returns of research focusing specifically on the impact of recent accountability systems, and especially their impact on secondary schools. One of the most careful and comprehensive studies at this level of education is reported by Carnoy, Elmore, and Siskin (2003). Their study found that high schools in four states differed significantly in responding to accountability systems, an important point that I will return to under lesson #3. Overall, however, schools reported only slight to modest gains in state standardized test scores, and these gains were restricted to states with high-stakes testing policies. Moreover, even such limited gains failed to correlate with any other measure of student achievement. In short, no corresponding gains were found in SAT scores, graduation rates, grades, or college admissions.

The finding that accountability testing does not predict any other educational outcome calls into question whether or not such tests tell us anything at all about the quality of a school or school district. Yet, because the state scores sometimes do go up, we should also ask another question: Exactly what is being learned? Are we simply preparing students to take a particular test? Or to borrow Tompkin's (2004) expression, we seem to be producing "little more than skilled test takers" (p. 80).

Studies that have examined the impact of accountability in actual classrooms tend to support this pessimistic interpretation. McNeil's (2004) research in Texas provides particularly vivid accounts of substantial class time spent bubbling in practice test answers and drilling students on commercial test preparation materials. For better or worse, students may be learning enduring lessons from these experiences. Yet, McNeil dubs the trivial pursuit that she found in so many classrooms a mandated "non-curriculum." The term non-curriculum aptly captures the meaningless rote learning that accountability seems to encourage. Increasingly, it seems that students find themselves learning skills that will not transfer to *anything* they will ever be asked to do in their lives outside of an examination room.

Earlier I argued that standards in and of themselves do not *cause* learning. It should also be evident from the research discussed above that testing in and of itself does not *cause* learning. On this point, educational policy makers might take counsel from a common saying among farmers: you cannot fatten livestock by putting them on a scale. But where does this leave us? Are standards and testing completely worthless? Not quite. Both are valuable to the extent that they inform rather than supersede the judgment of classroom teachers, and the possibility of using standards to inform teaching leads into the next lesson.

Lesson #2: Standards cannot replace judgment. This lesson focuses on a second limitation of standards as educational tools. It may at first seem counterintuitive, but recognizing the limitations of standards increases the likelihood of realizing their value, while ignoring their limitations or expecting too much threatens actual harm to the educational process. Educators and the public today risk ignoring such limitations when they mistakenly use the terms standards and goals interchangeably. It is right to associate values with goals, but not with standards. The difference between values and standards is significant. John Dewey (1934) emphasized this distinction more than seventy years ago. In one of his major works, Dewey wrote that:

> There are three characteristics of a standard. It is a particular physical thing existing under specified physical conditions; it is *not* a value. The yard is a yard-stick, and the meter is a bar deposited in Paris. In the second place, standards are measures of definite things, of lengths, weights, capacities. The things measured are not values, although it is of great social value to be able to measure them, since the properties of things in the way of size, volume, weight, are important for commercial exchange. Finally, as standards of measure, standards define things with respect to *quantity*. To be able to measure quantities is a great aid to further judgments, but it is not itself a mode of judgment [emphases in the original]. (p. 307)

Under increasingly difficult conditions of work, some teachers may understandably be tempted to substitute standards for their own professional judgments when it comes to either the content or the methods of their teaching. In this case, standards may give teachers "one less thing to worry about." Yet the test prep "non-curriculum" cited earlier points to the dangers of ignoring Dewey's point that standards can inform but not decide matters of value. Not only are such materials mind-numbing for students, but they undermine the autonomy and rewards of teaching. Citing a range of evidence, Wood (2004) wrote that:

> Teachers across the map complain that the joy is being drained from teaching as their work is reduced to passing out worksheets and drilling children as if they were in dog obedience school. Elementary "test prep" classroom methods involve teachers snapping their fingers at children to get responses, following scripted lessons where they simply recite prompts for students or have children read nonsense books, devoid of plot or meaning. (p. 39)

When standards replace judgments, simply teaching to the test places teachers in a double bind. While the teacher's autonomy and authority are reduced, their blame is not. On the contrary, for some, accountability is about shifting blame for school failure from inadequate support to poor teaching.

Lesson #3: Accountability cannot solve the achievement gap. One of the most attractive promises of NCLB, and one of the main reasons the law garnered bipartisan support when originally passed, was the hope that accountability measures could help lessen the stark inequities in American education by calling attention to the sometimes dismal education offered low-income children. In particular, early arguments pointed to the achievement gap between white and minority students as evidence that accountability was sorely needed. Accountability came to be touted as a way to equalize achievement. Today, however, NCLB and most state accountability systems face the disappointing irony of having the opposite effect. NCLB does target low-income schools, but it targets them for blame rather than for resources, thereby disadvantaging those students who are already the worst off.

Even without the serious limitations discussed above, accountability is unlikely to serve as an effective means to level the playing field. The inequities of American education are simply too extreme. Darling-Hammond (2004) notes that "the wealthiest U.S. public schools spend at least ten times more than the poorest schools—ranging from over $30,000 per pupil at the wealthy schools to only $3,000 at the poorest" (p. 6). Overcoming such huge differences would be difficult to achieve simply by insisting that poor schools try harder.

This realistic view is supported by several studies. In particular, the Carnoy et al. (2003) study cited earlier identified low-income schools that were performing poorly before the implementation of state accountability systems. Carnoy and his associates labeled this sub-sample in their study the "target schools" because these schools, according to the equity argument, were the schools most likely to benefit from accountability. Rather than benefit, however, these schools responded to accountability by following a pattern that DeBray, Parson, and Avila (2003) call "compliance without capacity" (p. 84). This research team suggested that low-income schools respond with superficial nods to accountability because the demands of state policies are eclipsed by the survival demands of keeping an inadequately funded school running day-to-day. Again referring specifically to the target schools, Elmore (2003) concludes: "These schools essentially did what they thought the law required with a minimum of alteration in their basic way of organizing and delivering instruction—and, not surprisingly, produced little in the way of improved performance" (p. 200).

I will return shortly to the issue of school capacity. Yet, the fact that some schools have far fewer resources than others is only half the story. Our expectations for the promise of accountability should also be tempered by recognizing that many, if not most, of the factors that researchers associate with the achievement gap are not under the school's direct control. Specifically, non-school related predictors of achievement include birth weight, nutrition, parent availability, television viewing patterns, and student mobility (Barton, 2004a; Freeman, 2005; Rothstein, 2004a). Accountability policies have not changed the social inequalities by which these factors vary across income and ethnicity.

For these reasons, educators and the public can hardly expect accountability to significantly reduce the achievement gap. Yet even more worrisome is the possibility that accountability policies may actually widen the gap. How might accountability disadvantage some schools over others? In the case of NCLB, schools that serve low-income students face a disadvantage because schools are declared failing if any one of their designated subgroups of students does not meet annual test goals. With more diverse ethnic populations and larger numbers of special needs students, low-income schools have a higher statistical probability of failing to meet "adequate yearly progress" (Darling-Hammond, 2004). This racially linked bias is unintended, but as Karp (2004) concludes, "The larger and more culturally diverse a school is, the more likely it is to be labeled as inadequate by NCLB" (p. 55).

As a consequence, the administrators at low-income schools probably have more motivation than other administrators to divert their already meager funds from more substantive programs related to curriculum to purchase expensive test-prep curricula. Another strategy is to emphasize tested subjects (particularly reading and math) over others. Siskin (2003b) reports the policy at one Texas high school that requires low-scoring students to simultaneously take one period of basic math, a second period of remedial math, and a third period of test prep math. "There is little time for other subjects," Siskin writes, "none for electives, and it is difficult to imagine a more perverse incentive for preparing well-educated high school graduates" (p. 180). Another study conducted by the Council for Basic Education (von Zastrow & Janc, 2004) surveyed 956 elementary and secondary schools to find that schools with high minority populations have been affected disproportionately by having to reduce their support for the arts, foreign language, and elementary social studies (pp. 7–8).

When student groups are targeted for intensive interventions, they risk more than just a narrow curriculum and being drilled for hours on content for which they have no interest. A California study (Novak & Fuller, 2003) recently found that failure to meet annual goals under NCLB was typically due to a single subgroup's low scores rather then to overall declining scores. What happens, then, when everyone knows that their school was labeled failing due to a particular minority group's scores, for example, or because of students with special needs? Will students understand the long and entrenched history of American educational inequities, or will they find it easier to blame and thus further stigmatize those fellow students they believe to be at fault?

Being labeled as a failing school may also influence teacher retention rates, which are yet another factor that researchers have linked with student achievement (Barton, 2004a). In particular, researchers who study teacher attrition worry about the effects of accountability on a school's ability to attract and retain experienced teachers. If given the choice, will teachers want to invest their careers in a school that has already been declared "failing"? Recognizing that attrition rates in education are already embarrassingly high, Ingersoll (2004) concludes that: "If the new accountability measures only serve to increase pressures on teachers without providing commensurate increases in their autonomy and resources, then they may end up simply driving even more teachers out of the occupation" (p. 146).

In a final bit of irony, states such as Maine and Maryland report abandoning sophisticated and hard-won performance assessment systems in order to make way for the testing requirements of NCLB (Berger, 2005; Darling-Hannond, 2004). Twenty years ago, evaluation experts were exploring performance assessment models as a more accurate way to assess the range of knowledge and skills that students acquire. It was hoped that such models would foster a greater sense of responsibility among teachers for the full range of what their students learn. It now seems, however, that performance assessment is but another casualty of NCLB.

Lesson #4: Teachers are accountable first to their students. Education takes a step backward if by making schools more accountable to state and federal politicians, we make them less accountable to their students. Is there a way to think about accountability that avoids this foolishness? My final lesson asks that we reaffirm a teacher's accountability to their students. Doing so again brings into focus the school-related factors that contribute to student achievement. Specifically, recognizing the social and economic sources of educational inequities points to the need for policies aimed at increasing school capacity, attracting and rewarding teachers who stay in the profession, assuring school safety, encouraging authentic assessments, reducing class size, and enriching the curriculum. Research tells us that these factors make a difference (Rothstein, 2004b), but attention seems to be easily diverted by what Linn (2000) aptly calls the "political quick fix" of test-driven accountability.

Reaffirming a teacher's accountability should also bring back into focus the relationships between teachers and students. If we could agree, even for a short time, that these relationships are more important than standardized test scores, we might ask what is likely to strengthen them and what is likely to weaken them. How accountable can high school teachers be, for example, when they see as many as 120 different students a day? And what happens to this relationship when students change teachers at the end of each twelve-week term?

Structural alternatives to traditional teaching assignments include block scheduling and multiyear teaching (see Flinders, 2000; Flinders & Noddings, 2001), the many variations of which schools should have an opportunity to explore. A given school, its staff, and its student body face unique circumstances, and for this reason, such alternatives should be weighed and discussed in local settings. The flexibility necessary for such deliberations may frustrate national policy makers and state Education Departments, especially those who were hoping for easy solutions to the always difficult challenges of educating for democracy. Still, we should continue to insist that education is at least as much about working directly with children and adolescents as it is about national "report cards."

Conclusion

I have argued that preparation for teaching should involve more than a familiarity with one's content and grade-level standards and where to find these standards online. Instead, future teachers should discuss openly and think critically about accountability and its unintended as well as intended consequences. I have also identified four lessons that I hope would facilitate this preparation. The first lesson is that standards do not cause student learning. The second lesson is that standards cannot replace the judgments of classroom teachers. The third lesson is that accountability alone cannot solve the achievement gap between white and minority students. On the contrary, accountability as it is presently conceived is likely to widen that gap. The fourth lesson is that teachers are accountable first to their students.

Ignoring the limitations and dangers of accountability will only distract future teachers from the important contributions that they are capable of making to the improvement of their schools and to the betterment of society at large. Teachers by themselves cannot solve poverty, homelessness, hunger, and the lack of affordable health care. However, they can make a difference in the lives of poor, homeless, hungry, and needy children. The degree to which teachers are helped to make this difference will be the degree to which they are genuinely accountable. Our own accountability and professional responsibilities as teacher educators should be to support teachers in this important work.

References

Barton, E. (2004a). Why does the gap persist? *Educational Leadership, 62*: 9–13.
Berger, Rebecca H. (2005). *Teacher capacity and assessment reform: Assumptions of policy, realities of practice.* Bloomington, IN: Indiana University Ph.D. dissertation.
Carnoy, M., Elmore, R., & Siskin, L. S. (Eds.) (2003). *The new accountability: High schools and high-stakes testing.* New York: RoutledgeFalmer.
Darling-Hammond, L. (2004). From "separate but equal" to "No Child Left Behind": The collision of new standards and old inequalities. In D. Meier & G. Wood (Eds.). *Many children left behind.* (pp. 3–32). Boston: Beacon Press.
DeBray, E., Parson, G., & Avila, S. (2003). Internal alignment and external pressure. In M. Carnoy, R. Elmore, & L. S. Siskin (Eds.). *The new accountability: High schools and high-stakes testing.* (pp. 58–85). New York: RoutledgeFalmer.
Dewey, J. (1934). *Art as experience.* New York: Pedigree Books.
Eisner, E. W. (1967). Educational objectives—help or hindrance? *School Review, 75*: 250–260.
Elmore, R. (2003). Accountability and capacity. In M. Carnoy, R. Elmore, & L. S. Siskin (Eds.). *The new accountability: High schools and high-stakes testing* (pp. 195–209). New York: RoutledgeFalmer.
Flinders, D. J. (Ed.) (2000). *Block scheduling: Restructuring the school day.* Bloomington, IN: Phi Delta Kappa International.
Flinders, D. J., & Noddings, N. (2001). *Multiyear teaching: The case for continuity.* Bloomington, IN: Phi Delta Kappa International.
Freeman, E. (2005). No Child Left Behind and the denigration of race. *Equity and Excellence in Education, 38*: 190–199.
Ingersoll, R. (2004). Revolving doors and leaky buckets. In C. Glickman (Ed.). *Letters to the next president* (pp. 141–147). New York: Teachers College Press.
Karp, S. (2004). NCLB's selective vision of equality: Some gaps count more than others. In D. Meier & G. Wood. (Eds.). *Many children left behind* (pp. 53–65). Boston: Beacon Press.
Linn, R. (2000). Assessments and accountability. *Educational Researcher, 25*: 4–16.
McNeil, L. (2004). Creating new inequalities: Contradictions of reform. In D. J. Flinders & S. J. Thornton (Eds.). *The Curriculum Studies Readers, Second Edition* (pp. 273–284). New York: RoutledgeFalmer.
Mager, R. F. (1962). *Preparing instructional objectives.* Palo Alto, CA: Fearon.
National Commission on Excellence in Education. 1983. *A nation at risk.* Washington, DC: U.S. Government Printing Office.
Noddings, N. (2003). *Happiness and education.* Cambridge: Cambridge University Press.

Novak, J. R., & Fuller, B. (2003). *Penalizing diverse schools? Policy brief 03-4*. Berkeley, CA: Policy Analysis for California Education.

Popham, J. (1970). Probing the validity of arguments against behavioral goals. In R. Kibler, L. Baker, & M. David (Eds.). *Behavioral objectives and instruction* (pp. 115–124). Boston: Allyn & Bacon.

Rothstein, R. (2004a). The achievement gap: A broader picture. *Educational Leadership, 62*: 40–45.

Rothstein, R. (2004b). *Class and schools*. New York: The Economic Policy Institute and Teachers College Press.

Siskin, L. S. (2003a). Outside the core. In M. Carnoy, R. Elmore, & L. S. Siskin (Eds.). *The new accountability: High schools and high-stakes testing* (pp. 87–98). New York: RoutledgeFalmer.

Siskin, L. S. (2003b). When an irresistible force meets an immovable object. In M. Carnoy, R. Elmore, & L. S. Siskin (Eds.). *The new accountability: High schools and high-stakes testing* (pp. 175–194). New York: RoutledgeFalmer.

Tompkins, Rachel. (2004). Eight million voices. In Carl Glickman (Ed.). *Letters to the next president* (pp. 77–84). New York: Teachers College Press.

von Zastrow, C., & Janc, H. (2004). *Academic atrophy: The condition of the liberal arts in America's public schools*. Washington, DC: Council of Basic Education.

Wood, G. (2004). A view from the field: NCLB's effects on classrooms and schools. In D. Meier & G. Wood (Eds.). *Many children left behind* (pp. 33–52). Boston: Beacon Press.

Summary and Implications

Michaelann Kelley
Eisenhower High School

Donna Reid
Educational Consultant

Gayle Curtis
Hamilton Middle School

Ron Venable
Eisenhower High School

P. Tim Martindell
Houston A+ Challenge

Allison Hamacher
Drew Academy for Mathematics, Science, and the Arts

Chapters 1 through 3 point out what amounts to three international trends in education: the demand for more and better forms of high-stakes testing, the associated de-professionalization of teaching, and the accompanying dehumanization of the education process. The first chapter by Wheatley and Rogers focused attention on the countless aspects of society that have been reduced to numbered representations. It contrasted measurement with feedback, asserting that the singular use of quantitative data for improvement instills disconnection and dispassion, whereas feedback intrinsically builds engagement, ownership, and the capacity for sustained change.

In chapter 2, Kelchtermans contrasted a business contractual relationship (performativity) with teacher self-understanding, which he believes is foundational for high-quality teaching and learning. Flinders, in chapter 3, confirmed that testing (high-stakes, high-standards, performativity, and measurement) is inherently necessary, but only as it provides opportunities for reflective practice in teaching and learning.

In the three chapters, readers find contrasting vocabularies and varying perspectives on a contentious issue that affects every educator in the U.S. and many other countries throughout the world. High-stakes testing has become established as a norm in schools, an unquestioned part of "the way things are." Each of the chapters attempts to provide new ways to view standards and how they could be integrated into school practices and curriculum.

Wheatley and Rogers, for example, suggest that promoting an organization's growth while sustaining teacher vitality requires moving past clerical number crunching to the internal development of measures and feedback directly related to the context of the organization. Drawing on his personal narrative authority, Kelchtermans links the concept of teachers' self-understanding to the individual teacher's ability to balance the internal and external locus of control with the characteristics that preservice teachers identify from their own memories of powerful teaching (Olson, 1995). Kelchtermans discusses the juxtaposition of the empowerment that accompanies teacher professional knowledge, passion, and self-understanding with the oversimplified logic behind performativity, which reduces good teaching to easily measurable standards.

For Flinders, mandated standards and teacher accountability are attempting to turn teachers into classroom "cookie-cutter facilitators." This constrained view of teachers' work suggests that many teachers do not have the opportunity to share their passion about what they teach or how they teach their students. It constrains their ability to teach independence of thought along with the opportunity to explore inquiry and reflection. At the same time, teachers risk teaching to the test, simply producing "little more than skilled test takers."

Taken together, these chapters awaken readers to how current educational systems appear to be falling short of promised goals and objectives. The shortfall appears detrimental to society as the most vulnerable citizens, youth, are being shortchanged in terms of their educational opportunities. At the same time, teachers who resist accountability pressures face penalties ranging from shunning, reassignment of position, and loss of employment. In sum, Wheatley and Rogers, Kelchtermans, and Flinders bring into public view the realities and complexities inherent in standards, standardization, accountability, and performativity that teachers encounter daily as they strive to educate the students they meet face-to-face in classrooms and schools.

Reference

Olson, M. (1995). Conceptualizing narrative authority: Implications for teacher education. *Teaching and Teacher Education, 11*(2), 119–135.

Division 2
IMPACT AND CONSEQUENCES OF ACCOUNTABILITY SYSTEMS

Overview and Framework

Susan McCormack
University of Houston–Clear Lake

Denise McDonald
University of Houston–Clear Lake

Tirupalavanam G. Ganesh
University of Houston

Andrea S. Foster
Sam Houston State University–Huntsville

>Susan McCormack, Ed.D., is an assistant professor of social studies education at the University of Houston–Clear Lake.
>
>Denise McDonald, Ed.D., is an assistant professor at the University of Houston–Clear Lake. She teaches undergraduate and graduate level courses on classroom management, curriculum planning, and instructional strategies.
>
>Tirupalavanam G. Ganesh, Ph.D., is a principal research analyst for the Office of the Dean at the Mary Lou Fulton College of Education at Arizona State University, Tempe.
>
>Andrea S. Foster, Ph.D., is an assistant professor in the curriculum and instruction department at Sam Houston State University in Huntsville, TX, where she teaches elementary and middle school science methods and graduate classes in testing and measurement.

The April 2006 editorial headline of a widely circulated urban newspaper that appeared on the eve of the mandated statewide administration of the Texas

Assessment of Academic Skills (TAKS) read, "TAKS dishonors the true educators!" The columnist took the position that ". . . there is something especially wrong with a test that teachers and students believe, sucks the energy and creativity out of education" (Clack, 2006, p. 01J). Too often the short stories of Flannery O'Connor, the plays of Sophocles, or the poetry of Neruda are put away to concentrate on preparing for the test. Would Albert Einstein or William Shakespeare be permitted to develop their artistic talents if they lived in our existing age of accountability?

Such wonders and ponders arising from real world situations positions us to ask the very important accountability question: Is it necessary to measure what students know and how well they have learned? If the answer to this question is in the affirmative, then we need to ask (as Wheatley and Rogers urged in chapter 1) why we need this information, what exactly will we measure, how the measurement will occur, and lastly, how this information should be used?

One can argue that without objective knowledge about how well our schools are functioning, we cannot strengthen our educational systems. Worldwide, measures of students' knowledge, skills, and abilities are often obtained through tests. These measures of student achievement are frequently used as indicators of how well schools are functioning—i.e., they are used to hold schools and the people who function within the schools accountable. Tests, when used appropriately, could be considered unassailable and unprejudiced measures of students' knowledge. However, "Professional judgment is the foundation for assessment and, as such, is needed to properly understand and use all aspects of assessment" (Rudner & Schafer, 2002, p. 7).

Nevertheless, when student achievement information as measured by tests is used as a single measure of performance for a student, a teacher, or a school, the adoption and implementation of tests can have detrimental consequences. Without a doubt, a single test cannot be valid for all purposes. Indeed, tests differ in their ability to provide significant measurements of student learning (Rudner & Schafer, 2002). Large scale efforts to measure student knowledge that are national and international in scope are important; however, equally important is the way in which the resulting information is used. Without care given to its use, information gained from measurements can have severe consequences—both intended and unintended—for the student, the teacher, the school, and the community.

In the United States, the National Assessment of Educational Progress (NAEP) has been billed as the nation's report card that allows for state-wide comparisons of student achievement. The NAEP is designed to test a representative sample of U.S. students in 4th, 8th, and 11th grades in reading, mathematics, and science on a 4-year cycle. Student achievement data collected through programs such as the Trends in International Mathematics and Science Study (TIMSS) and the Programme for International Student Assessment (PISA) permit comparison of student achievement across nations. The TIMSS provides data on the mathematics and science achievement of U.S. students compared to that of students in other countries. The PISA is an internationally standardized assessment administered to 15-year-olds in schools in over 40 countries to assess some of the knowledge and skills that are essential for full participation in society. With PISA, students who are at the end of compulsory or mandated K–12 education are tested in reading, mathematical, and scientific literacy.

While the NAEP data have revealed gaps in achievement by race and ethnicity within the U.S., data from administrations of the TIMSS and PISA have revealed that U.S. children are performing well below the median in international comparisons. Arguments about U.S. public school quality and the need to hold the nation's public schools "accountable" are often supported with NAEP, TIMSS, and PISA data (Hanushek, 2005).

The call for accountability already evident in many U.S. states prior to 2001 was enhanced by the sweeping legislation of the No Child Left Behind Act of 2001 (NCLB). By 2003, NCLB had an impact on school accountability systems in every state. The nature of the impact and the intended and unintended outcomes of related federal and local school accountability laws continue to be heavily debated (Berliner, 2004; Hanushek, 2005; Hanushek & Raymond, 2004; Nichols, Glass, & Berliner, 2006; Rothstein, 2005; Smith, 2004). Nevertheless, political agendas and school reform efforts focused on accountability have shaped "high-stakes" testing in the U.S.

What makes a test a high-stakes test? Policy makers mandate a test to collect information about student achievement and use the test results to attach consequences for students, teachers, administrators, schools, and school systems. The mandated administration of these tests and the ensuing student achievement information are subsequently used to hold public school systems accountable. The consequences attached to test results have included student promotion or retention, high school graduation, salary increases for teachers and educators, autonomy to operate schools, school funding, school choice, and attachment of achievement labels to schools. When educational decisions are made based on test-scores, the tests become high-stakes (Swope & Miner, 2000).

The impact and consequences of high-stakes testing on educational systems are all-encompassing and are discussed in this section of the *Yearbook*. The chapter authors focus on consequences of testing results that range from professional development for teachers to student motivation for completion of mandatory schooling and aspirations for life after high school.

Fletcher, Strong, and Villar assess the effectiveness of new teacher support programs in three California districts on their students' reading achievement using the required data from the yearly testing program (Stanford 9). Attempts to show the impact of new teacher support programs on student achievement are indeed complex. Such studies require attention to several factors related to new teacher induction and mentoring including mentor to novice teacher ratio; mentor teacher quality and experience; the preparation necessary for mentors to play the role of a mentor; availability of mentors; and the quality, length, and frequency of mentoring activities. Other necessary considerations are attention to specific aspects of student characteristics and demographics, the curriculum, and implementation of the curriculum.

Perhaps, as with any information resulting from program evaluation efforts, what really matters is how ensuing program decisions impact the lived lives of the individuals closest to the testing phenomenon—the students, the teachers, and their school communities. When funding determinations, decisions on continuance, or changes to programs are made based on test results, the high-stakes nature of testing is significantly magnified.

Lee examines the effect that high-stakes tests such as high school exit exams and college entrance tests have on students' cognitive and affective outcomes. Taking an international focus, Lee investigates 8th-grade mathematics assessment and related survey data from the Trends in International Mathematics and Science Study (TIMSS) for Japan, Korea, England, and the U.S. Careful consideration of social, cultural, and academic contexts of educational systems and the nature of the TIMSS survey questionnaire is crucial in drawing inferences from international comparisons of student achievement. Do students in Japan and Korea express aspirations for higher education beyond high school more so than students in England and the U.S.? How does the prevalent national high-stakes setting impact students' interest in academic and social life beyond high school? Apart from school support systems for student achievement, societal expectations for academic knowledge, the national and local economy, and the job market are important aspects to consider in understanding student outcomes for K–12 education.

Klug suggests that standardized tests do not take into account all factors of academic success. Whereas middle-class students have shown increased academic performance over the past 30 years, the same is not true for students in lower socio-economic regions, which represents many Native American students. As a result, Native American students perceive educators to be uncaring and preoccupied with culturally irrelevant practices. Klug suggests that improvements are possible when teachers use a variety of strategies that are culturally sensitive and support the local community. While many Native American populations enjoy dual citizenship, individual heritage is constantly attacked by arbitrary guidelines and historical governmental interference. To Klug, adherence to NCLB policies does little more than reaffirm historical federal assimilation processes.

Snow-Gerono focuses on the change in teaching and learning perceptions.

Mentor teachers in elementary schools candidly regard the impact of testing on the classroom environment and contend that educators cannot sit idly while others re-create their profession: Teachers must become active meaning makers. High-stakes testing impacts the preservice teacher experience as well. Job satisfaction is diminishing and reveals itself to inexperienced teachers who are rapidly losing interest in a profession that is mandated by outside influences. Teacher education programs and local districts face a huge dilemma—how to balance theory with practice and continue to build strong partnerships between the university and neighboring school districts as they come together to negotiate meaningful learning at all levels.

A. Ganesh explores standardizations' impact and influence on teachers' pedagogy and students' schooling experiences through interviews and observations. Collectively, the teacher participants believe that children learn at different paces and that instruction should be individualized. They expressed how the standardized testing focus places a distinct effect on what they teach, how they teach, and instructional timeframes in which they teach. One teacher in this study shared the impact of standardization on relational pedagogy with respect to her efforts in developing how students feel about themselves as learners and how they combat a feeling of insecurity.

References

Berliner, D. C. (2004). If the underlying premise for No Child Left Behind is false, how can that act solve our problems? In K. Goodman, P. Shannon, Y. Goodman, & R. Rapoport (Eds.), *Saving our schools* (pp. 167–184). Berkeley, CA: RDR Books.

Clack, C. (2006, April 16). TAKS dishonors true educators, *San Antonio Express*, p. 01J.

Hanushek, E. A. (2005). Why the federal government should be involved in school accountability. *Journal of Policy Analysis and Management, 24*(1), 168–172.

Hanushek, E. A., & Raymond, M. E. (2004). Does school accountability lead to improved student performance? (Working Paper 10591). Cambridge, MA: National Bureau of Economic Research. Retrieved May 20, 2005, from http://papers.nber.org/papers/w10591.

Nichols, S. L., Glass, G. V., & Berliner, D. C. (2006). High-stakes testing and student achievement: Does accountability pressure increase student learning? *Education Policy Analysis Archives, 14*(1). Retrieved January 4, 2006, from http://epaa.asu.edu/epaa/v14n1/.

Rothstein, R. (2005). Why the federal government should not be involved in school accountability. *Journal of Policy Analysis and Management, 24*(1), 172–177.

Rudner, L. M., & Schafer, W. D. (2002) (Eds.). *What teachers need to know about assessment*. Washington, DC: National Education Association.

Smith, M. L. (2004). *Political spectacle and the fate of American schools*. New York: Routledge Falmer.

Swope, K., & Miner, B. (2000). *Failing our kids: Why the testing craze won't fix our schools*. Milwaukee, WI: Rethinking Schools.

CHAPTER 4

Accountability Systems and Program Evaluation

AN INVESTIGATION OF THE EFFECTS OF VARIATIONS IN MENTOR-BASED INDUCTION ON THE PERFORMANCE OF STUDENTS IN CALIFORNIA

Stephen Fletcher
New Teacher Center, University of California, Santa Cruz

Michael Strong
New Teacher Center, University of California, Santa Cruz

Anthony Villar
New Teacher Center, University of California, Santa Cruz

> Stephen Fletcher, Ph.D., is a researcher with the New Teacher Center at the University of California, Santa Cruz. His research interests focus on the relationship between student achievement and teacher professional development. He is also interested in cross-organizational collaboration to improve teacher training.
>
> Michael Strong, Ph.D., is the research director at the New Teacher Center at the University of California, Santa Cruz. Dr. Strong is a recognized expert in the relationship between new teacher support and teacher retention. He is also interested in the nature of mentor-novice conversations.
>
> Anthony Villar, MPP, is a researcher at the New Teacher Center at the University of California, Santa Cruz. His research interests include mentor development and educational policy.

ABSTRACT

> The current emphasis on accountability systems may provide educators with the opportunity to evaluate the effectiveness of instruc-

tional programs. Using data from a state's annual testing program, researchers analyzed achievement gains of classes taught by new teachers as one means of evaluating new teacher support in three districts. The results provide useful information about the relationship between program support for new teachers and student learning. Thus, a silver lining to the accountability cloud may be new information about the effectiveness of professional development.

There is a growing concern about the quality of education in the United States. Long-term trend data from the National Assessment of Education Progress (NAEP) indicates that less than half of 17-year-old students are reading at a level that allows them to understand the information they are learning in school. NAEP results in mathematics indicate that 40% of 17-year-old students have difficulty computing decimals, simple fractions, and commonly encountered percents. The results of the Third International Mathematics and Science Study (TIMSS) led researchers to describe the United States curriculum as "a mile wide and an inch deep" (Schmidt et al., 1997). Thus, large-scale studies of education lead people to question whether students are learning and what is being taught in classrooms.

The large-scale studies of education provide valuable information at a national level, but they do not provide parents with information about their children's school or other local districts. To address the issue of local schools and districts, Congress modified the Elementary and Secondary Schools Act to strengthen the accountability of state and local educational agencies in order to provide more information to parents/guardians. As enacted, the No Child Left Behind (NCLB) legislation (2001) requires that students in grades three through eight be tested each year in reading and mathematics and that the results be used as indicators of school quality. Thus, parents now have access to student achievement information about their local school/district and other districts in their state.

Despite the intentions of legislators, accountability systems associated with NCLB have been criticized by the education community. The criticisms have centered on the accountability systems being based on student testing. One group of critics has questioned the requirement that the majority of students in grades three through eight be tested. The questions have centered on when limited English students should be tested, and whether special education students should participate (Abedi, 2004). Another group of critics has questioned accountability systems from the perspective of how the assessments will effect instruction. Smith (1991) found that the high stakes testing in Arizona led some teachers to stop traditional instruction several weeks prior to testing in order to spend time on test preparation strategies.

A shortcoming of the criticisms of accountability systems is that they have limited the discussion among educators on possible advantages of NCLB. One advantage may be that students in grades three through eight are tested annually in English/language arts and mathematics. For some states, these assessments are based on state-specific curriculum standards which have been agreed upon by educators, politicians, and parents. Other states have purchased a norm-referenced assessment which has been reviewed by the state board of education and the public. Yearly testing means last year's test results can be compared with this year's results to determine what students have learned. This information, then, can be part of the data used to evaluate the effectiveness of instructional programs.

A type of program that could be evaluated using data from yearly student testing is new teacher support. New teacher support is the process by which novices are initiated into the teaching profession and into a particular school. New teacher support can vary from a brief orientation about district policies and procedures to working with a mentor for one or more years. The American Federation of Teachers (2001) argued for more extensive support of new teachers when they said, "Beginning teachers need the support, advice, and guidance that only experienced teachers can provide" (p. 3).

Smith and Ingersoll (2004) found that 83% of new teachers receive some degree of mentor support. They also found that the amount of support given to new teachers is related to the likelihood of their staying in their school and in education. For example, if the only support received was the opportunity to work with a mentor, 40% of new teachers either changed schools or left the profession. If new teachers work with a mentor, receive assistance from a site administrator, participate in a network of beginning teachers, and receive a reduced class load, then 90% stay in their school or the profession. However, Smith and Ingersoll did not have the data to determine if the amount of support received by new teachers helped them to be effective in teaching their students.

The purpose of this study is to illustrate how the annual testing required by accountability systems can be used to study instructional programs. Specifically, the study looks at the change in the achievement of classes taught by new teachers in different districts. Thus, the study provides information about the effectiveness of new teacher support programs, as well as illustrating the value of accountability systems.

Method

Three school districts in California were asked to participate in the study. The districts were in various stages of implementing a formative assessment system for their induction program. Two of the districts, A and B, chose to implement the formative assessment system in the context of their existing mentoring program. District C had used the formative assessment system for several years. Therefore, the study was seen as a way to compare variations in new teacher support, with District C most closely resembling Smith and Ingersoll's intensive program.

District Demographics

As shown in table 4.1, the three districts varied with respect to enrollment, the percentage of students receiving free/reduced cost lunch, student achievement, and the percentage of Latino students. District A had the smallest percentage of minority students and students receiving free/reduced cost lunches. Districts B and C are similar in terms of Latino enrollment, but District B has a higher percentage of students receiving free/reduced cost lunches.

Classes taught by new teachers also varied demographically across districts. As indicated in Table 4.2, classes taught by new teachers were similar in class size but varied in terms of student poverty and student ethnicity. Comparing tables 4.1 and 4.2 indicate that classes taught by new teachers have more minority students and poor students than the district average.

District Induction Programs

The new teacher support programs in all of the districts last for two years. As indicated in table 4.3, the mentor-novice ratio ranged from 1:12 to 1:15 for

Table 4.1 2002 demographic statistics for three districts using mentor-based induction

District	Total enrollment	% Latino	% free/reduced lunch	Average total reading score (in NCEs) by grade level				
				2	3	4	5	6
A	51,383	26.1	27.3	55.9	55.9	57.5	55.4	56.4
B	81,058	80.9	74.8	41.9	39.0	39.6	39.0	42.5
C	19,863	78.2	53.5	39.6	38.3	40.1	40.7	41.3

Table 4.2 Characteristics of classes taught by novice teachers

	District A	District B	District C
Class size	23	25	25
Average total reading score before the start of school	52	34	32
Percentage of students that is a minority	27%	87%	87%
Percentage of students receiving free/reduced cost lunch	42%	60%	100%
Intensity of induction program	1.25	1.75	2.00
Number of students in the study	424	709	1288
Number of teachers in the study	17	31	51

mentors working with first-year teachers. The mentor-novice ratio in District A goes to 1:1 for second-year teachers because they are mentored by a colleague in the same school. The mentor-novice ratio for second-year teachers in district B is 1:32. The mentor-novice ratio for second-year teachers in district C remains at 1:15.

The information in table 4.3 is the basis for the Induction Intensity variable in table 4.2. The values for each district's program are based on four components: (a) whether the program involved a mentor, (b) how selective the program was in recruiting mentors, (c) the amount of professional development for mentors, and (d) the contact time between mentor and novice. A district was assigned a quarter-point for each year its program involved a component. Selectivity of mentors and the amount of professional development for mentors are estimated by the number of mentors required by the mentor-novice ratio. For example, site-based mentors (1:1 ratio) means more mentors need to be hired and trained, so selectivity of mentors and the amount of professional development may be reduced. For the three districts, each program received one point for Year 1 because each program used all four components. In Year 2, district A had one component (mentors), district B had three components (mentors, selectivity, professional support for mentors), and district C had each of the four

4.3 Characteristics of the mentor-based induction programs for three districts

District	Time for induction	Mentor classification		Mentor:novice ratio	
		Year 1	Year 2	Year 1	Year 2
A	2 Years Release	Full Release	No (on site)	1:12	1:1
B	2 Years	Full Release	Full Release	1:12	1:32
C	2 Years	Full Release	Full Release	1:15	1:15

components. Consequently, values for the induction variable were 1.25, 1.75, and 2 for districts A, B, and C, respectively.

In addition to the information in table 4.3, it is important to note that each of the districts used the same reading curriculum in the elementary grades. This factor is critical because if districts were using different curriculum, then any difference in achievement gains could be attributed to curriculum and not teacher support. Thus, the use of the same curriculum across districts limits alternative explanations to the findings.

Student Achievement

We obtained student and teacher data associated with the spring 2001 and spring 2002 administrations of the Stanford Achievement Test, version 9 (SAT/9) from the three districts. Student achievement was defined using the Total Reading score (in Normal Curve Equivalents).

According to table 4.1, district A had the highest initial achievement scores, with an average Total Reading score of 55, while districts B and C had an average score of 40. Table 4.2 indicates that classes taught by new teachers in district A had an average score of 52, classes in district B had an average score of 34, and district C classes had an average of 32. It is apparent that new teachers are generally assigned classes that are below the district average in terms of reading achievement.

Analysis

Student achievement data were analyzed using a hierarchical linear model with student information as level-1 and teacher and class information as level-2. The level-1 equation was:

$Y_{ij} = \beta_{0k} + \gamma_{ij}(\text{Pre-Class Achievement}) + \delta_{ij}(\text{Minority Status}) + \epsilon_{ij}$.

The equation assumes that the amount of student learning in a school year depends on what a student knows at the beginning of the school year and certain demographic characteristics (Ballou, 2002; Kupermintz, 2003). The Pre-Class Reading data were re-centered using the class mean.

The level-2 equations for the coefficients of the level-1 variables were determined using exploratory features of the HLM software. Potential level-2 variables were selected based on the existing literature (Hanushek et al., 2002). The level-2 variables include: (a) proportion of students in a class that received free/

reduced cost meals (Class Poverty), (b) proportion of students in a class that were minority (Class Minority), (c) the average SAT/9 Total Reading score of students in the class before the start of the school year (Pre-Class Reading), and (d) an induction variable (Induction) defined in terms of the four components of the intensity of the program (mentor, mentor selectivity, support for mentors, mentor-novice ratio). Based on data from the exploratory analysis, variables were added to the level-2 equations until the preliminary t statistic was less than 1. Thus, the inclusion of variables in the level-2 database was guided by the extant literature, but inclusion of variables in the level-2 equations was guided by preliminary results of the HLM analysis.

Results

As indicated in table 4.4, four models were tested in the analysis. Model 1 is the simplest of the models with only Induction and Class Minority included in the level-2 equations. Model 4 is the most elaborate model as it includes Induction Intensity, Class Minority, and Class Poverty. One thing that is common across the models is that the coefficient associated with Class Poverty is always negative, regardless of how the level-2 variable is added to the analysis. The other negative coefficient in the analysis is associated with the interaction of Student Minority Status and Class Minority. Finally, Model 1 accounts for the highest percentage of variance (73.99) of the models, but the other models do account for 70% of the variance.

Based on the principle of parsimony, we will focus the discussion on Model 1. The results of Model 1 indicate that some of the level-2 variables have a direct influence on student achievement and some influence student achievement through interaction with student-level variables. For the level-1 Intercept, Induction is statistically significant in the level-2 equation (23.918, $p<.01$). This result indicates that induction for a teacher has a direct positive effect with student achievement. The results also indicate that the initial achievement level of a class (Pre-Class Reading) has a positive interaction with Induction (.326, $p<.01$) and Class Minority Status (.157, $p<.05$). The results also indicate that Class Minority Status has a negative interaction with Student Minority Status (-4.659, $p<.01$).

As the coefficients have not been standardized, it is useful to do some estimates to illustrate which variables make the greatest contribution to the outcome. The Induction variable ranges from one to two for the three districts in the study, so we will use 1.5 to illustrate a moderately intensive induction program. For classes taught by new teachers in the three districts, the average Class Minority variable is .8 and the Class Poverty variable is .7. For class achieve-

Table 4.4 Models of student achievement using student and class variables

Level-1 (student) variable	Level-2 (class) variable	Model 1	Model 2	Model 3	Model 4
Intercept	Intercept induction		32.057**	38.197**	38.138**
	Class poverty	23.918**	−17.998**	−31.588**	−31.718**
Preclass reading	Intercept induction			.539**	.497**
	Class minority	.326**	.349**		.161*
	Class poverty	.157*	.098		
Student minority status	Intercept				
	Class size				
	Class minority	−4.659**	−4.835**	−5.032**	−4.667**
	Preclass achievement				
% variance		73.99	70.88	70.07	70.08

** $p < .01$
* $p < .05$

ment, we can use a value of 32, which is close to the achievement level of classes in Districts B and C. If we define our test case as a minority student with a Pre-Class Reading score of 40, we find that the majority of the predicted achievement is based on the Induction variable associated with the level-1 Intercept. The value of this estimate is that it indicates supporting new teachers can have a positive influence on student achievement, regardless of the poverty or minority status of a class.

A different way to evaluate the results of the analysis is in terms of program evaluation. The simplest test is to look at the predicted score for a minority student (Student Minority Status = 1) in a poor, minority classroom (Class Minority = 1.0, Class Poverty = 1.0) whose pre-class reading score matched the class average (Pre-Class Reading = 0). The induction values are 1.5 for moderate intensity (e.g., site-based mentor who receives support) and 2.0 for high intensity (full-release mentor receiving support). The predicted score for the student taught by a new teacher supported by a high intensity program indicates a gain of 9–11 points in student achievement. For a moderately intensive program, the model predicts a student will lose one to two points. Consequently, high intensity new teacher support is related to positive gains in student achievement, and a moderate intensive support is associated with negative gains.

Discussion

The purpose of this study is to explore how the new emphasis on accountability associated with the No Child Left Behind legislation can be beneficial to educators. We argue that the yearly testing in grades three through eight required by the legislation provides an opportunity to analyze professional development programs for new teachers. To illustrate, we use student achievement data from three districts in California to compare the effectiveness of their programs supporting new teachers. The results of the study indicate that analyzing student test data can provide insights into which components of support may be of most benefit to improving student learning.

Another value of this study is that it provides some guidelines about what factors need to occur in order for testing data to be used for accountability. For example, the analysis used in this study required being able to match student information across years. This requirement means that students need to be assigned a unique identification number when they enter the district, and this identification number stays with them until graduation or exiting. Many districts already assign unique student identification numbers, which indicates the districts can do longitudinal analysis of student achievement. Other districts will need to begin with this step.

Another factor that needs to be in place for districts to benefit from the emphasis on accountability is an adequate research and evaluation department. As we have contacted districts in California, we have learned that the latest budget cuts have led to staff reduction in district-level research departments. Some responsibilities of the departments have shifted to other district administrators, but these people had a high enough workload for fulltime employment before the new duties. It is also possible the people assigned new duties do not have the research or evaluation background to do the work. Intermediate-level education agencies (e.g., county offices of education) may be able to provide some support, but their assistance will be constrained by staff training and how many school districts need help. It is possible the analytical assistance may be provided by faculty at colleges and universities who are seeking research opportunities. Consequently, educational organizations may need to re-think their relationships in order to take advantage of accountability data.

The use of accountability data is not without controversy, though. The main criticism to the current emphasis on accountability has focused on testing and how testing has limited the school curriculum (Smith, 1991). This criticism has value in some cases but not others. For example, the criticism was applicable to the California's Standardized Testing and Reporting (STAR) program from its inception until 2004 because of the focus on norm-referenced tests. The argument lost its strength, though, when the state accountability system shifted from norm-referenced testing to criterion-referenced tests designed in accordance with state curriculum standards. Thus, the narrowing of the school curriculum in California, and other states, may be more related to the development of content standards, not testing practices.

Based on this study, the No Child Left Behind legislation may provide some added opportunities for educators to learn about programs for students and teachers. Without the districts trying to implement the assessment program in grades three through eight, this study would not have been possible. Without a concern about the effectiveness of programs, the districts may not have been willing to participate in this study. Thus, the emphasis on accountability may give school districts the opportunity to study programs supporting teachers as well as students.

References

Abedi, J. (2004). The No Child Left Behind Act and English language learners: Assessment and accountability issues. *Educational Researcher, 33*(1), 4–14.

Ballou, D. (2002). Sizing up test scores. *Education Next, 2*, 10–15.

Kupermintz, H. (2003). Teacher effects and teacher effectiveness: A validity investigation of the Tennessee value added assessment system. *Educational Evaluation and Policy Analysis, 25*(3), 287–298.

No Child Left Behind Act of 2001, House of Representatives, 107 Sess.(2001).

Schmidt, W. H., McKnight, C., & Raizen, S. A. (1997). *A splintered vision: An investigation of U.S. Science and mathematics education.* Boston: Kluwer Academic Publishers.

Smith, M. L. (1991). Put to the test: The effects of external testing on teachers. *Educational Researcher, 20*(5), 8–11.

Smith, T. M., & Ingersoll, R. M. (2004). What are the effects of induction and mentoring on beginning teacher turnover? *American Educational Research Journal, 41*(3), 681–714.

CHAPTER 5

Revisiting the Impact of High-Stakes Testing on Student Outcomes from an International Perspective

Jaekyung Lee
State University of New York at Buffalo

>Jaekyung Lee, Ph.D., is an associate professor of education at the State University of New York at Buffalo. His research focuses on educational accountability and equity. His publications include "The Impact of Accountability on Racial and Socioeconomic Equity" in the *American Educational Research Journal*, 2004 (with K. Wong), and "Racial and Ethnic Achievement Gap Trends" in *Educational Researcher*, 2002.

ABSTRACT

>The impact of high school exit exams and college entrance depends on whether the tests matter to student success and thus influence their learning efforts and outcomes. This paper revisits evidence on the impact of high-stakes testing on cognitive versus affective student outcomes from an international perspective with focus on the cases of Japan, Korea, England, and the U.S. It provides a critical look into the role of high-stakes testing in a broader social and cultural context of educational systems based on the analysis of the Third International Mathematics and Science Study (TIMSS) 8th-grade mathematics assessment/survey data. The mixed results suggest that the impact of high-stakes testing varies between the Eastern and Western countries and that students' aspiration for high school graduation and higher education may act as a moderator of testing effects.

Introduction

Tests have grown into instruments of educational accountability with increasing control over educational systems in many countries (Eckstein & Noah, 1993). Indeed, performance-based accountability was the hallmark of education reform in many countries during the last two decades, and high-stakes testing (HST in short hereafter) has become the linchpin of this policy movement. The global educational accountability movement seems to have been shaped by growing international competition in the context of the global economy and related concerns about standards and quality in education (Kearns & Doyle, 1991; Husén & Tuijnman, 1994).

It has been argued that the United States is one of the rare developed countries without a high-stakes exit exam for high school students and thus American students do not work as hard as their counterparts in higher-performing countries (Bishop, 2001; Tomlinson & Cross, 1991). With increasing concerns about poor student performance and school accountability in the U.S., increasing numbers of states adopted HST, such as high school exit exams, during the 1990s. Further, the *No Child Left Behind Act* (NCLB) of 2001 mandates statewide testing with serious consequences attached to the test results for schools.

However, HST and policies of test-driven external accountability in American school systems are relatively recent, and there is a lack of a strong empirical knowledge base to guide policy decisions on these issues. In light of this problem, should American policy in testing and accountability practices be modeled after what high-performing countries do?

In the midst of an international brain race, people were often led to believe that students in the countries ranking highest on international achievement tests (e.g., Japan, Korea, and Singapore) have a better education. Indeed, American and British educational policymakers attempted to model policies such as national curricula and testing on those of higher-performing Asian countries (Lee, 2001). There are some drawbacks in such international comparisons and benchmarking efforts based on test scores. Comparative researchers have revealed mixed findings on different types of student outcomes in high-performing vs. low-performing countries. Asian students, who outperform American students, tend to have lower self-concepts of ability (Stigler, Smith, & Mao, 1985). Whang and Hancock (1994) attributed this inconsistency to high standards in Asian countries that make students evaluate a good performance as being not quite good enough. Asian students' relatively poor academic motivation and attitudes were also revealed by results from the Third International Math and Science Study (TIMSS).

Given international variations in different types of student outcomes, countries that participated in TIMSS may provide a laboratory for investigating the relationship between national testing policies and student outcomes. We need to understand how HST affects cognitive versus affective student outcomes and also how the social and cultural context of education moderates the impact of testing on student outcomes. This paper attempts to fill the gap by investigating international variations in testing policy, school context, and student outcomes.

For cross-cultural comparison, this study focuses on four developed countries—two selected from the East (Japan and Korea) and two selected from the West (England and the United States). Japan and Korea may typify the Eastern developed countries that have highly centralized school systems and uniformly high educational aspirations. In contrast, the United States and England may represent the Western developed countries where educational governance is decentralized and educational aspirations are relatively diverse.

Previous studies showed cross-cultural differences between Asian and American students and their parents regarding aspirations and expectations for academic achievement and educational attainment (see Holloway et al., 1990; Lee, 2005; Stevenson & Stigler, 1992). From this cross-cultural comparative perspective, this study explores possible influences of educational aspiration on student outcomes in response to HST.

Mixed Evidence and Debates about HST

International comparative studies have found substantial differences in the requirement, nature, and type of exams for students who complete secondary education. Countries that have a national high school exit exam or college entrance exam still vary in terms of the purpose, rigor, and competitiveness of the tests (Eckstein & Noah, 1993). Bishop (1998) examined whether there were curriculum-based external high school exit exams (CBEEES) among different TIMSS countries and how the policy variations are related to different student math and science achievement outcomes. The study showed significant positive effects of high school exit exams on achievement variables.

On the contrary, reanalysis of the same data by Baker and his colleagues (Baker, Akiba, LeTendre, & Wiseman, 2001) challenged the original finding by showing an insignificant effect of Bishop's CBEEES variable. Recent reanalysis of Bishop's CBEEES with larger student-level data sets of TIMSS and TIMSS-repeat by Wößmann, L. (2003) showed significantly positive effect of central exit exams.

Within the United States, the case that drew the most attention was in Texas, where the evidence on the effects of HST on achievement was mixed and often contradictory (Carnoy, Loeb, & Smith, 2001; Haney, 2000; Skrla, Scheurich, Johnson, & Koschoreck, 2004; Valencia, Valenzuela, Sloan, & Foley,

2004). While there were many studies that compared states' test-driven external accountability policy in relation to academic achievement, the analyses of these studies have also produced inconsistent findings (Amrein & Berliner, 2002; Grissmer & Flanagan, 1998; Carnoy & Loeb, 2002; Raymond & Hanushek, 2003; Lee & Wong, 2004).

Previous studies have also examined the impact of HST and accountability policies on students' affective outcomes. Some showed a positive impact of policy on academic motivation and behaviors, including improvement in students' course-taking (Shiller & Muller, 2003), student motivation (Roderick & Engel, 2001), and teacher motivation (Kelley, Heneman, & Milanowski, 2000). On the other hand, other studies showed negative impacts such as increased levels of anxiety, stress, and fatigue among students as well as decreased morale among teachers (Jones, Jones, Hardin, Chapman, Yarbrough, & Davis, 1999; Abrams, 2004). It was argued that high-stakes tests can decrease student motivation to learn and lead to higher student retention and dropout rates (Amrein & Berliner, 2003).

International Variation in HST Policy

A high-stakes test is a test that has serious consequences for students' future education and social mobility, such as promotion, graduation, or entrance into a college or university. The meaning and importance of HST for high school students is likely to vary from country to country, depending on social definitions of success and on opportunities available for students to achieve defined success in each country. Students who finish their secondary school with success typically pass some examination, receive a diploma, and/or are rewarded for their achievement with a place, scholarship, or fellowship in higher education (Eckstein & Noah, 1993).

Among the 39 countries that participated in the 1995 TIMSS, 22 countries, including England, Japan, and Korea, were classified by Bishop (1998) as having CBEEES in both math and science. Four had CBEEES in math only; 5 had some CBEEES but not all parts of the country; and 8 had none. The U.S. is one of the countries that had CBEEES in parts of the country, that is, in some states but not in others.

Bishop's analysis is limited in capturing the effects of HST. First, it excludes university entrance examinations by arguing that such exams should have much smaller incentive effects because they only target college-bound high school students, thus, schools and teachers can avoid responsibility for test results. However, this reasoning does not hold true in countries that have universal aspiration for postsecondary education and excessive competition for college entrance examinations. Secondly, there was misclassification of Japan and Korea as having a national high school exit exam since both have a university entrance exam instead of an exit exam.

In this chapter, I also use 1995 TIMSS 8th-grade data and reclassify the TIMSS countries by broadening the definition of HST to include both high school exit exams and college entrance exams. This analysis also has several limitations. Firstly, although high school data may better capture the effects of HST on achievement, it only examines 8th-grade TIMSS results. Secondly, dichotomous classification of countries based on the requirement of a high school exit exam or college entrance exam does not differentiate countries in terms of the level of stakes for students. Finally, it focuses on the type of HST that targets students but excludes other types of testing and accountability policies that have more direct consequences for schools and teachers.

In England, Japan, Korea, and the United States, admission into a prestigious university is the standard signal of high success, but there are different selection and competition processes (see table 5.1). In England and Japan, there are pre-selection mechanisms in place.

In England, the General Certificate of Secondary Education (GCSE) exam at age fifteen or sixteen determines who will continue into upper secondary school and have the opportunity to study for the General Certificate of Education Advanced Level (GCE A-level) examination. In Japan, entrance examinations are administered for placement in senior high schools at age fifteen, and those who are selected into top-tier senior high schools have better chances of going to prestigious universities. Later, applicants for university admission go through two stages of university entrance examinations: (1) the Test of the National Center for University Entrance Exam (TNCUEE) and (2) an entrance exam administered by individual universities.

Similar to Japan, Korea has a national system of university entrance exam, the College Scholastic Ability Test (CSAT), but the role of testing for selection is relatively weak. In the U.S., there is no universal national-level exam for high school completion or university entrance. Instead, a high school exit exam is only required in some states, and the requirement and importance of nongovernmental college entrance exams such as the SAT and ACT varies among the different types of colleges and universities.

International Variations in Student Outcomes

HST may cause undue pressure for students to pass tests for graduation and/or to win limited places at prestigious colleges. HST at the high school level may

Table 5.1 High school exit exams and college entrance exams in Korea, Japan, England, and the United States

Country	Name and type of exam	Who controls?	Who takes and when?	What are the consequences?
Korea	Entrance exam for upper secondary school	Prefecture (regional/district level)	All students seeking admission to high school	Placement in high school track (academic vs. vocational)
	College Scholastic Ability Test (CSAT)	Nation	All students seeking university admission	Selection and placement in university
Japan	Entrance exam for upper secondary school	Prefecture (regional/district level)	All students seeking admission to high school	Placement in high school
	Test of the National Center for University Entrance Exam (TNCUEE)	National	All students seeking university admission	Selection and placement in university
	Entrance exam for individual universities	Individual universities	Students seeking admission to specific universities	Selection and placement in university
U.S.	University entrance exams—SAT, ACT	Nongovernmental	College-bound students in last 2 years of secondary school	Varies by college/university (most 4-year colleges require exam, most 2-year colleges do not)
	High school exit exam (minimum competency tests or end-of-course exams)	States (administered in some states)	High school students in last two to three years	Varies by state (most states allow for repeat testing to pass)
England	General Certificate of Secondary Education (GCSE)	National	All upper secondary school students at age 16	Preselection device for a second national exam and ranking schools for accountability
	General Certificate of Education Advanced Level (A level) and Advanced Supplementary (AS level) for entrance to postsecondary institutions	National	All students seeking university admission. Typically taken at age 18.	Major factor in higher education placement

Sources: Adapted from *Education Indicators: An International Perspective* (NCES 96-003). Washington, D.C.: NCES, 1996.

have trickle-down effects on students at the middle school level to improve their learning efforts and achievement outcomes. Figure 5.1 shows some support for this hypothesis since countries with HST tend to outperform those without it on average in 8th-grade mathematics. However, this generalization is not applicable in all cases. Among the four case study countries, Korean and Japanese students perform at the top while British and American students perform below the international average.

Excessive competition among students in this academic race may result in detrimental effects on students' attitudes such as self-concept of learning ability and motivation for learning. The comparison of countries with HST versus countries without HST provides some support for this hypothesis as the former group tends to have less positive affective outcomes than does the latter. However, the patterns are not always consistent (see figures 5.2 and 5.3).

Korean and Japanese students tend to report the lowest level of academic self-concept of mathematics achievement among all the TIMSS countries, while American and British students are placed in the top ranks. This perceived level of students' math performance is sharply contrasted with actual math achievement level as measured by the TIMSS standardized test. Further, students in Japan and Korea also report relatively weak interests in mathematics compared with their counterparts in England and the U.S.: only 53 percent of Japanese and 58 percent of Korean students reported liking math, whereas 71 percent of American students and 80 percent of English students reported liking math.

The Impact of Testing Policy and Context on Student Outcomes

The above comparisons of TIMSS countries' HST policy and student outcomes between Japan, Korea, England, and the United States raise questions about the role of social and cultural contexts that may moderate the effect of testing on outcomes. The underlying assumption of HST is that the tests matter very much to students for their success and thus will influence their learning efforts and outcomes. Does the assumption apply equally well to students in every country? Perceived level of stakes attached to tests is likely to vary among countries according to the level of their students' educational and occupational aspiration. In this study, it is hypothesized that the consequences of high school exit exams and/or college entrance exams for student outcomes depend on underlying incentives for doing well on the tests, which reflect how much students aspire to complete secondary education and seek entrance into higher education.

Some countries like Japan and Korea are well known for parents' exception-

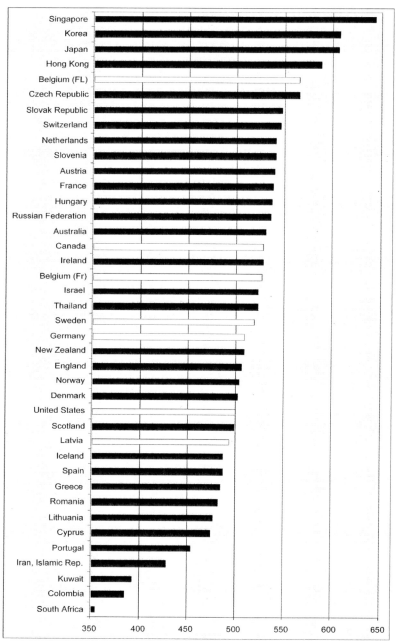

Figure 5.1 National average mathematics scores of 1995 TIMSS 8th graders (black bars for countries with HST)

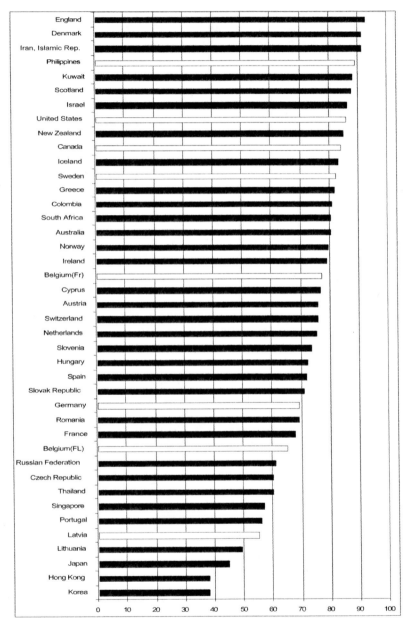

Figure 5.2 Percentages of 1995 TIMSS 8th graders who agree or strongly agree that they usually do well in mathematics (black bars for countries with HST)

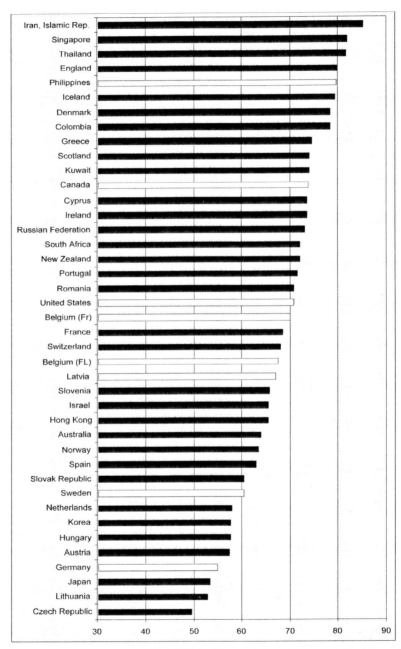

Figure 5.3 Percentages of 1995 TIMSS 8th graders who report that they like mathematics or like math a lot (black bars for countries with HST)

ally high aspiration for their children's educational attainment and achievement. Indeed, Korea and Japan rank at the top among all the TIMSS countries regarding 8th-grade students' aspiration for completion of high school and postsecondary education (see figure 5.4). American students are behind those two countries but they fare well. No comparative information is available for England due to its missing data for this TIMSS student survey question.

This study uses multiple regression to explore the effects of national HST on three measures of student outcomes, including average math achievement test scores, self-concept of math ability, and attitudes toward math (see Appendix). Three related social and cultural context variables that may confound the relationship between high stakes testing and student outcome variables are included in this regression for statistical control: The TIMSS countries' income level (measured by GNP), East Asia (proxy for regional cultural values), and educational aspiration (measured by the percentage of students who reported that they are planning to complete high school and beyond). The results of separate regressions of three outcome variables are reported in table 5.2.

GNP and higher education aspiration rate show significant positive effects on math performance (*beta* $= .32, p < .01$ for GNP; *beta* $= .43, p < .01$ for educational aspiration). Students in East Asian countries also tend to have a higher math performance on average (*beta* $= .36, p < .01$). Once these factors are held constant, countries that have high school exit exams do not perform significantly better or worse than do their counterparts without such exams (*beta* $= -.01, p > .05$).

While there is no significant effect of HST alone for any type of student outcome, the joint effect of HST and educational aspiration (as shown by the interaction between the two variables) is statistically significant, implying that the effect of HST may be moderated by educational aspiration. However, the direction of such interactive effect tends to vary by the type of outcome: positive for cognitive outcome (*beta* $= .25, p < .05$ for math achievement) and negative for affective outcome (*beta* $= -.35, p < .05$ for self-concept of math; *beta* $= -.46, p < .05$ for attitudes toward math). High school exit exams and/or college entrance exams may boost student achievement when students have strong motivation for completion of high school and seek entrance into higher education. But, at the same time, the exams may have detrimental effects on students' self-concept and attitudes when students feel incompetent or stressed as a result of excessive competition to pass or excel at the exams.

Conclusion

HST is prevalent in many countries, as it has grown as a key policy instrument to improve academic achievement. Notwithstanding global trends of the HST

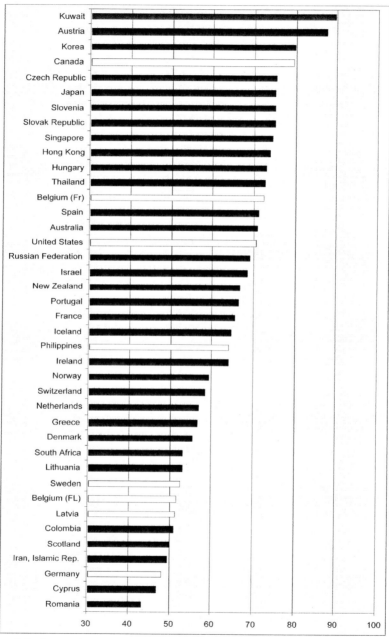

Figure 5.4 Percentages of 1995 TIMSS 8th graders who aspire to finish secondary school or college/university (black bars for countries with HST)

5.2 Results of multiple regression of TIMSS 8th-grade math student outcomes on predictors including high-stakes testing (HST), educational aspiration, East Asia, and GNP

	Dependent variables		
	Math achievement (test scores)	Self-concept of math achievement	Attitudes toward math (liking math)
High-stakes testing (HST)	−1.24	−6.22	−1.78
	−.01	−.18	−.08
Educational aspiration	1.88**	−.16	−.28
	.43	−.12	−.34
East Asia	50.14**	−17.27**	6.95
	.36	−.45	.29
GNP	.001**	.000	−.000
	.32	.10	−.10
Interaction of HST	2.56*	−1.07	−.89
and aspiration	.25	−.35	−.46
constant	369.07***	88.09***	84.123***
R^2	.66	.43	.29
N	36	37	37

Note: Unstandardized regression coefficients appear above standardized coefficients (beta weights are italic). Asterisks indicate the level of statistical significance for the coefficients: * $p < .05$, ** $p < .01$, and *** $p < .001$.

and test-driven educational accountability movement, we need to understand international variations in testing policies, contexts, and outcomes. For example, there is no high school exit exam in Japan and Korea, but instead they both have national college entrance exams that impact only college-bound students. However, in Korea and Japan a majority of students have aspirations for higher education and experience strong competition for top-tier universities. As such, the college entrance exam may have substantial effects on average student outcomes. These students show relatively higher achievement and lower self-concept of math ability and liking of math in general. In contrast, England where national educational aspiration is relatively lower, in keeping with the tradition of elite higher education, the impact of nationwide HST policy on student outcomes would be felt less.

The contrast of national testing policies and student outcomes between the East and West reveals the tension between improvement of academic knowledge/skills and attitudes under HST. HST may have helped students accomplish higher academic achievement at the expense of their academic self-concept and attitudes toward learning. This study calls for more balanced attention to both cognitive and affective outcomes for an integrated evaluation of HST effects.

Educational policymakers need to revisit the goal of accountability policy from a whole-person education perspective.

The findings of this study suggest that HST for high school students such as a high school exit exam and/or a college entrance exam may affect student outcomes only if there are strong incentives for students to perform well on the tests. With strong incentives for students, HST may function as a double-edged sword by producing either a positive or a negative impact. When students pursue high school graduation and/or college education, HST as a gatekeeper in conjunction with high academic standards may work to improve students' learning efforts and achievement. On the other hand, excessive academic pressure due to HST and challenging standards may hurt students' self-concept of ability and attitudes toward learning, particularly in the absence of social support.

In the U.S. and many other industrial countries, a changing economy has caused the high school diploma to serve as a minimum requirement for increasingly more jobs and has increased the currency of a college education. Within this context, pressures for doing well on high school exit exams and/or college entrance exams have been raised. This change also elevates needs for increased support and assistance for students and their schools who are held accountable for the test results. Educational policy and practice for HST need to build upon the fact that underlying expectations, incentives, and support for better student outcomes come from all educational institutions and social institutions beyond schools. Understanding the connection between secondary education and the job market, and the interface between secondary and postsecondary education is crucial.

Appendix: Description of Variables

Math Achievement (Test Scores). This variable measures 8th-grade students' average math achievement level based on the TIMSS standardized math assessment results. The TIMSS math proficiency scores are scaled to have a mean of 500 and a standard deviation of 100.

Self-Concept of Math Achievement. This variable is derived from the TIMSS student survey statement: "I usually do well in mathematics at school." From the 4-point scale ranging from "strongly disagree" to "strongly agree," the percentage of 8th-grade students in each country who chose "agree" or "strongly agree" is reported as an index of students' confidence in doing school math.

Attitudes toward Math (Liking Math). This variable is derived from the TIMSS student survey statement: "I like mathematics." From the 4-point scale ranging from "dislike a lot" to "like a lot," the percentage of 8th grade students in each country who chose "like a lot" or "like" is used as an index of students' positive attitudes toward math.

Educational Aspiration. This variable is derived from the TIMSS student survey question about one's plan for further education: "How far do you expect to go?" The percentage of 8th-grade students in each country who reported "complete secondary school, some postsecondary education, or finish university" is used as an index of educational aspiration.

High-Stakes Testing (HST). This is a dummy variable indicating whether each country has a high school exit exam and/or college entrance exam with consequences for students. It is based on several sources including Bishop (1998), NCES (1996), and Final Report of Education Conference Communiqué (1999). Countries that have such exams only in parts of their region are treated as having no national exams.

GNP. This variable is gross national product per capita as reported in the UNESCO's (1997) Statistical Yearbook. It is an index of each nation's economic development.

References

Abrams, L. M. (2004). *Teachers' views on high-stakes testing: Implications for the classroom.* (Education Policy Studies Laboratory Policy Brief) Tempe, AZ: EPSL.

Amrein, A. L., & Berliner, D. C. (2003). The effects of high-stakes testing on student motivation and learning. *Educational Leadership 60,* 5, 32–38.

Baker, D. P., Akiba, M., LeTendre, G. K., & Wiseman, A. W. (2001). Worldwide shadow education: Outside-school learning, institutional quality of schooling, and cross-national mathematics achievement. *Educational Evaluation and Policy Analysis 23, 1,* 1–17.

Bishop, J. (2001). A steeper, better road to graduation. *Education Next.* Retrieved October 6, 2005, from www.educationnext.org/20014/index.html.

Bishop, J. (1998). *Do curriculum-based external exit exam systems enhance student achievement?* (CPRE research report RR-40) Philadelphia, PA: CPRE.

Carnoy, M., Loeb, S., & Smith, T. (2001). *Do higher scores in Texas make for better high school outcomes?* (CPRE Research Report No. RR-047) Philadelphia, PA: Consortium for Policy Research in Education.

Carnoy, M., & Loeb, S. (2002). Does external accountability affect student outcomes? *Educational Evaluation and Policy Analysis, 24*(4), 305–331.

Eckstein, M. A., & Noah, H. J. (1993). *Secondary school examinations: International perspectives on policies and practice.* New Haven: Yale University Press.

Education Conference Communique (1999). *Examination and certification system at the end of the upper secondary general education.* Final report of the Education Conference Communique. Downloaded on October 19, 2005, from www.mszs.si/exam/final.html.

Grissmer, D., & Flanagan, A. (1998). *Exploring rapid achievement gains in North Carolina and Texas.* Washington, DC: National Education Goals Panel.

Haney, W. (2000). The myth of the Texas miracle in education. *Educational Policy Analysis Archives.* Retrieved March 3, 2001, from http://epaa.asu.edu/epaa/v8n41.

Holloway, S. D., et al. (1990). The family's influence on achievement in Japan and the United States. *Comparative Education Review, 34,* 196–208.

Husén, T., & Tuijnman, A. (1994). Monitoring standards in education: Why and how it came about. In A. Tuijnman, & T. N. Postlethwaite (Eds.). *Monitoring the standards of education* (pp. 1–21). Oxford, U.K.: Pergamon.

Jones, G., Jones, B., Hardin, B., Chapman, L., Yarbrough, T., & Davis, M. (1999). The impacts of high-stakes testing on teachers and students in North Carolina. *Phi Delta Kappan, 81*(3), 199–203.

Kearns, D. T., & Doyle, P. D. (1991). *Winning the brain race.* San Francisco: ICS Press.

Kelley, C., Heneman, H., & Milanowski, A. (2000). *School-based performance award programs, teacher motivation, and school performance: Findings from a study of three programs* (CPRE Research Report RR-44) Philadelphia, PA: Consortium for Policy Research in Education.

Lee, J. (2001). School Reform Initiatives as Balancing Acts: Policy Variation and Educational Convergence among Japan, Korea, England and the United States. *Educational Policy Analysis Archives.* Retrieved on October 10, 2005, from http://epaa.asu.edu/epaa/v9n13.html.

Lee, J. (Ed.). (2005). Revisiting Education Fever from International Perspectives [Special Issue]. *KEDI Journal of Educational Policy, 2*(1).

Lee, J., & Wong, K. K. (2004). The impact of accountability on racial and socioeconomic equity: Considering both school resources and achievement outcomes. *American Educational Research Journal, 41,* 797–832.

Leithwood, K., Steinbach, R., & Jantzi, D. (2002). School leadership and teachers' motivation to implement school accountability policies. *Educational Administration Quarterly, 38*(1), 94–119.

National Center for Education Statistics. (1996). *Education Indicators: An International Perspective.* NCES 96-003. Washington, DC: National Center for Education Statistics.

National Center for Education Statistics. (1996). *Pursuing excellence: A study of U.S. eighth-grade mathematics and science teaching, learning, curriculum, and achievement in international context.* Washington, D.C.: U.S. Department of Education, National Center for Education Statistics.

National Center for Education Statistics. (2000). *Pursuing excellence: Comparisons of international eighth-grade mathematics and science achievement from a U.S. perspective, 1995 and 1999.* Washington, D.C.: U.S. Department of Education, National Center for Education Statistics.

Raymond, M. E., & Haushek, E. A. (2003). High-stakes research. *Education Next,* Summer/No.3. Retrieved January 20, 2004, from www.educationnext.org/20033/index.html.

Roderick, M., & Engel, M. (2001). The grasshopper and the ant: Motivational responses of low-achieving students to high-stakes testing. *Educational Evaluation and Policy Analysis, 23*(3), 197–227.

Rothstein, R. (2001, May 16). Weighing students' skills and attitudes. *The New York Times*, p. A25.

Shiller, K., & Mueller, C. (2003). Raising the bar and equity? Effects of state high school graduation requirements and accountability policies on students' mathematics course taking. *Educational Evaluation and Policy Analysis, 25*(3), 299–318.

Skrla, L., Scheurich, J. J., Johnson, J. F., & Koschoreck, J. W. (2004). Accountability for equity: Can state policy leverage social justice? In L. Skrla & J. J. Scheurich (Eds.) *Educational Equity and Accountability: Paradigms, policies, and politics* (Ch. 5, pp. 51–78). New York: RoutledgeFalmer.

Stevenson, H. & Stigler, J. (1992). *The learning gap: Why our schools are failing and what we can learn from Japanese and Chinese education.* New York: Summit Books.

Stigler, J. W., Smith, S., & Mao, L. W. (1985). The self-perception of competence by Chinese children. *Child Development, 56,* 1259–1270.

Tomlinson, T. M., & Cross, C. T. (1991). Student effort: The key to higher standards. *Educational Leadership, 49*(1), 69–73.

Valencia, R. R., Valenzuela, A., Sloan, K., & Foley, D. (2004). Let's treat the cause, not the symptoms: Equity and accountability in Texas revisited. In L. Skrla & J. J. Scheurich (Eds.) *Educational Equity and Accountability: Paradigms, policies, and politics* (Ch. 3, pp. 29–38). New York: RoutledgeFalmer.

Whang, P. A., & Hancock, G. R. (1994). Motivation and mathematics achievement: Comparisons between Asian-American and Non-Asian students. *Contemporary Educational Psychology, 19,* 302–322.

Wößmann, L. (2003). Central exit exams and student achievement: International evidence. In P. E. Peterson & M. R. West (Eds.), *No Child Left Behind? The Politics and Practice of School Accountability* (pp. 292–323). Washington, DC: Brookings.

CHAPTER 6

Broken Promises, High Stakes, and Consequences for Native Americans

Beverly J. Klug
Idaho State University

> Beverly J. Klug, Ed.D., is an associate professor of education in the Department of Educational Learning and Development, Idaho State University. She is the Idaho Chancellor, International Association of Educators for World Peace. Her research revolves around issues of American Indian education, multicultural education, social justice, literacy, and teacher education.

ABSTRACT

> Treaties between tribal nations and the U.S. federal government included provisions for the education of Native children in exchange for relinquishment of traditional lands. There is a long history of the federal government determining the type of education provided for Native students, primarily with the goal of assimilating students to the dominant Western culture. Consequently, many American Indians left an educational system that did not take into account their strengths or their needs. Promises made in recent history for Native educational self-determination have produced successful school programs; yet, recent legislation and concomitant regulations may destroy dearly gained culturally relevant programs.

High-stakes testing has had enormous effects on the education of American Indian students in regions throughout the country. The good news is that, because scores must be disaggregated by ethnicity for reporting to state and national entities (No Child Left Behind Act, 2001), attention has been drawn to the education provided for Native American students (the terms "American Indian" and "Native American" are used interchangeably as either is preferred in different areas of the nation). Awareness of schools that are performing at low levels and the problems of struggling students are public knowledge. Unfortunately, while teachers and schools must be accountable for students' actual learning, everyone is painted with the same brush when schools are seen as failing, regardless of the reasons.

This is one of the difficulties brought about through the accountability system required by the No Child Left Behind Act (2001), commonly referred to as NCLB. Punitive consequences await students, teachers, and schools that fail to make Adequate Yearly Progress for all students. Instead of celebrating gains made by individual students and innovative teaching leading to success, we are treated to the specter of sanctions if there is not compliance determined by test scores for entire school enrollees.

The scores of students attending private schools, whether church-based, home-school, tribal schools, or elite boarding schools, are not counted for reporting at state or national levels although these students take the assessments used in their states. Therefore, public opinion concerning the abilities of Native students and other underrepresented populations may be erroneous, or inaccurate, simply because of the reporting requirements of NCLB (2001).

Therein lies the dilemma: teaching and learning should be occurring in every school in the nation. How we measure both is the challenge. Added to this is the bad news that, because of the assessment re-norming requirement of NCLB (2001), many students will continue to lag behind even though their overall scores may be equal to the "proficient" levels of those reported early in the implementation of this legislation. Consequently, those who do not have an understanding of statistical procedures may draw conclusions that many Native students, and those of other ethnicities or disabilities, will always fall short of "average" for the country as a whole.

Who Are American Indians?

Native Americans enjoy dual citizenship within their sovereign Nations and the United States government. There are 513 federally recognized tribal Nations in the United States, and several others that are state-recognized (St. Germaine, 1995). During the period from 1945–1965, many other tribal Nations lost their Sovereign Nation status through the process of termination by the federal government. Many are petitioning to be recognized once again (Sheffield,

1997). The American Indian and Alaskan Native populations (reported as one category) are growing and together represent 1.5% of the population in the United States at the present time (Public Information Office, 2004).

Misunderstandings abound concerning qualifications for American Indian status (Sheffield, 1997). The federal government early on used blood as a determiner of Native American ethnicity. Different tribal Nations require specific amounts of blood quantum for enrollment purposes, which involves issuing a Certificate of Degree of Indian Blood (CDIB) certifying members' lineages (Snipp, 2000). The majority of tribal Nations, rancherias (California), or pueblos (American Southwest), require at least ¼ blood quantum (having one grandparent who is American Indian), though for some Nations enrollment depends upon parentage or other factors; consideration of qualifications may also vary from government agency to agency (Gonzales, 2001). The Cherokee Nation has the most liberal requirements requiring that ancestry can be traced to someone listed on the government's Dawes Rolls (Bureau of Indian Affairs, 1998).

When people marry across tribal affiliations, their children may not have the required blood quantum for enrollment in any Nation, even though they have only American Indian lineages. Some are unable to claim their heritages because tribal Nation regulations require that their mother or father belong to the Nation, and their parents may reflect the opposite, e.g., having a father from a matrilineal Nation (Sheffield, 1997).

In urban areas, it is much more difficult to determine Native American status due to the experiences of many American Indians who left their tribal Nations after World War II as part of the federal government's Relocation Program of the Bureau of Indian Affairs (Frazier, 1993). The program was designed to encourage assimilation and aimed to better economic conditions for Native families who lived in cities. Wax (1971) explains that many American Indians in urban areas denied their ethnicities due to the high rates of prejudice and discrimination they encountered on their job sites as well as in other life realms. They did not want their children to endure the same types of problems resulting from racism and stereotyping and felt that by not declaring their heritages, they were protecting their children in the future. Census figures indicate 60% of American Indians and Alaskan Natives live in urban areas today (Public Information Office, 2004).

As a result, the question of who is an American Indian is not an easy one to answer. Educators need this understanding since policies based on ethnicity may not actually reflect the reality of "who is and who isn't" and conclusions drawn vis-à-vis particular groups of people. Similarly, Native American students may be considered to be assimilated, bi-cultural, or traditional, dependent upon the way they are being raised within their families and communities. Assimilated students appear to relate more to the school's curriculum and practices. However, for all Native students, their academic success depends upon teachers' understandings of the interrelationships of the families, schools, and communities in which they teach (Bronfenbrenner, 1979).

The History of Education of American Indians in the United States

The federal government guaranteed education for all American Indians as a result of ceding land through treaties with tribal Nations (Kappler, 1904). The complexity of this provision revolves around what is considered "appropriate education." The earliest perspective taken by the government was to provide assimilationist education for Native students, training boys to be farmers and teaching girls industrial arts, with the intent of "civilizing" them (Nabovkov, 1978). Schools were organized around these themes, and eventually a boarding school system evolved, developed along military lines following the example of the Carlisle Indian School in Pennsylvania directed by Col. Richard Henry Pratt (Spring, 2004). School programming with active intent of eliminating Native American languages and cultures became the norm for American Indian education.

For some, the boarding schools offered opportunities not available otherwise. For the majority of students enrolled, the experience was devastating and severely impacted Native peoples (Johansen, 2000; Macqueen, 2000). Children were forbidden to speak their languages or practice their cultural traditions; many were removed from their families for years at a time. When and if they returned home, many had difficulties living in their communities, as they were no longer "Indian" in the way they thought or acted. Later, they were also unprepared to teach their languages, traditional cultures, and mores to their own children and grandchildren.

Racism, harsh punishment, physical, emotional, and other abuse that took place in the schools all left a devastating legacy still impacting Native communities today. The Meriam Report (1928/1977) detailed abuses taking place within the schools and made recommendations to establish community day schools for American Indians, including reviving Native languages and cultures. However, the damage done has had lasting effects on views of "White man's education" within Native communities (Skinner, 1992). The proviso that Native languages and cultures should be taught in schools has been largely ignored, as the majority of teachers have been non-Native and unfamiliar with the languages and cultures of the students they teach (Lomawaima & McCarty, 2002). The curriculum and materials used in public schools primarily reflect a dominant culture perspective that does not distinguish Native peoples for their strengths, but, if addressed at all, as the dominant culture negatively portrayed them in historical media.

Culturally Relevant Pedagogy

The graduation rate for American Indian and Alaska Native students (reported in aggregate) stands at 71% according to the latest census, with higher rates reported for biracial students (Public Information Office, 2004). Data from many sources inform us that the format of education for American Indian students fails to offer bicultural and traditional Native students' opportunities for success. Dropout rates reach highs of 40–60% in some parts of the country. Even though the rationale for NCLB (2001) is to stem the tide of dropouts nationally, in effect the act does not encourage more school participation in a system that does not recognize the unique strengths and needs of American Indian students.

The Indian Education Act of 1972 (U.S.C. 2000) and The Indian Self-Determination and Education Assistance Act of 1975 (U.S.C. 2000) were legislated in response to Native communities' desires for more ownership in the education of their children and called for a culturally sensitive public education (St. Germaine, 2000). Culturally relevant pedagogy requires that teachers understand and respect: (a) the cultural expectations of communities in the way they teach their children, show respect, and care for their people; (b) the traditional wisdom that has shaped communities; (c) the language and language praxis in students' communities; and (d) the importance of relationship to everyone and all that surrounds us (Klug & Whitfield, 2003).

These "ways of knowing" for Native students include long periods of observation, performance trials (practice of the skills) by oneself, demonstration to an adult of content mastery, and working in cooperative groups (St. Germaine, 2000). This type of pedagogy requires that teachers are willing to teach from both a dominant culture point of view and that of the tribal Nations they are serving while given the resources and support to do so (Pewewardy, 1994).

In essence, culturally relevant pedagogy does not separate children from what they know but uses their understandings to build new knowledge. Doing this not only produces deeper learning and more involvement in schools, it supports the constructivist philosophies of learning advocated by Vygotsky (1978), Dewey (1902), and Piaget (Wadsworth, 1996). From research about how the brain learns and processes information, it is now understood that concrete experiences engaging the perceptual areas of our brains are needed for learning to take place (Zull, 2002). These experiences must be reflected upon, hypothesized about, and actively tested.

Incorporating the arts that are intrinsic to Native cultures across content areas adds a significant dimension to learning by capitalizing on body-brain connections. In addition, it is imperative to provide educational experiences that are emotionally positive for learning to occur (Zull, 2002). Culturally relevant pedagogy reflects children's Native cultures and the ways they have been taught at home and in the community, thereby fostering an emergence of positive feelings about schooling and reducing student stress, which can inhibit the learning process.

Various tribal Nations developed information that could be utilized by teachers and students to learn about their histories and worldviews in the late 1980s and early 1990s (Strom, 1994). The *Alaska Standards for Culturally Responsive Schools* (Alaska Native Knowledge Network, 1998) provides an effective model for education of Alaska Native and American Indian students, outlining conditions for their success, including maintenance of languages and culture.

THE IMPACT OF NCLB (2001) ON EFFORTS TO SUSTAIN CULTURALLY RELEVANT PEDAGOGY

Lomawaima and McCarty (2002) detail the habitual struggles for control over the education of American Indian students between tribal Nations and the federal government since the inception of the United States. The current state of affairs in education reflects those struggles due to policies resulting from NCLB (2001). For instance, schools serving Native students located within larger school systems receiving federal funding may be forced to adopt one of the basal reading programs recommended by the National Reading Panel (2000). These programs are heavily phonics-based and for students who speak their Indian languages as their first language, or a dialect variance of English based on their Native languages (Leap, 1997), difficulties may arise with understanding and completion of required activities. The programs do not reflect pedagogy more conducive to Native student learning.

The assessments required to satisfy NCLB (2001) do not include Native languages and cultural concerns. Teachers in schools serving Native populations may find themselves continually stressed as they try to meet the government accountability demands and, therefore, do not have time to address cultural issues in the curriculum. Knowing that students' achievement scores are considered to reflect teaching abilities, educators may abandon what they have learned about teaching Native students in favor of preparing them for standardized assessments. There is presently no indication that the requirements for addressing culturally relevant pedagogy will become reflected in state or national assessments, negating previous legislation for American Indian education.

ACHIEVEMENT TESTING: THE GREAT DEBATE

Glasser (1998) states that many students are not enamored with schools, resulting in low performance whether they are capable of doing more or not. As a result, teachers resort to using coercive methods with students such as low grades or the threat of failure. No Child Left Behind (2001) reflects this same mentality. According to Glasser (1998), "What we lack is not more information from tests or any other source. It is the will to abandon our traditional 'teach, test, rank, and coerce the losers' system of education, which, at best, does not work for more than half the students" (p. 74).

While the causal aspects of differences in achievement for underrepresented populations are continually debated, we do know the language used for these tests is that of White, middle-class Americans and incorporates concepts retrieved from a dominant culture curriculum with which these children are usually familiar (Bernol, 2002). Abedi (2004) states that NCLB's (2001) conception of academic progress does not reflect all the factors involved in academic success, as in the case of the wide variance of characteristics of students designated as Limited English Proficient (LEP).

Berliner (2004) finds that there have been continuous gains in content areas evidenced by standardized assessment results over the last thirty years, primarily realized by middle-class children attending middle-class schools. Children from low-income areas, as many Native students on reservations or living in urban areas of poverty are, must deal with social and economic problems affecting their learning that the majority of middle-class students do not. When students perceive schools as unfriendly or punitive places, many will act out or act as if they are invisible (Testerman, 1996). These factors, combined with feelings and perceptions that teachers and administrators do not care about them, influence decisions many students make to drop out of school.

By focusing only on outcomes of learning, nurturing students takes a backseat to achievement tests. In many instances, teachers are told that they must prepare for the tests by practicing test-taking strategies with their students and spending hours covering material they believe will be on the tests. Because tests must be administered more frequently due to NCLB (2001), even more time is lost to true instructional activities than before this legislation was enacted.

An additional concern about standardized assessment measures is the presence of bias (Glasser, 1998). Most assessments given follow a multiple choice format due to ease in scoring and do not reflect the learning preferences of different groups of students. The assessments only measure a narrow range of material, not taking into account life differences, regional differences, and cultural understandings of the world. Fragmentation of information into pieces and parts presents its own difficulties for Native students who view the world holistically (Klug & Whitfield, 2003).

An alternative to making decisions about American Indian students' abilities to perform as reflected on standardized assessments would be to put more emphasis on authentic assessments created to measure what students are actually able to do and how they use the knowledge they have learned. In other words, focus needs to be on process knowledge, not just on the retrieval of information. Authentic assessments as recommended in *Indian Nations at Risk: An Educational Strategy for Action* (Indian Nations at Risk Task Force, 1991) are seen as highly congruent with traditional ways of teaching and learning in Native communities. These types of assessments have been recommended for many years within educational communities as a whole (Vygotsky, 1978; Glasser, 1998).

DO ACHIEVEMENT TESTS MEASURE LEARNING OR RETRIEVAL OF INFORMATION?

In the last decade, much has been learned about the brain and knowledge acquisition through advances in technologies such as Magnetic Resonance Imaging (MRI) and Computed Tomography (CT) scans (Morgan, 2004). These technologies have shown that different areas of the brain are activated at different times and for different tasks, and information is frequently processed in multiple areas simultaneously. It is estimated that humans are born with 100 billion neurons capable of making connections with as many as 10,000 other neurons (Zull, 2002). Making these connections, called "neuronal networks," is vital to the process of learning and remembering information.

According to Zull (2002), all learning has a neurobiological basis. As connections are made among the major areas of the brain known as the frontal lobes, temporal lobes, parietal lobes, occipital lobes, cerebellum, limbic system, and the cerebral cortex, what we learn becomes more strongly ingrained and retained. Responsibility for short-term memory, working memory, and long-term memory is then distributed to different systems throughout the brain.

PROCESS KNOWLEDGE VS. PRODUCT KNOWLEDGE

While the basis of learning is beyond the purview of this paper, what is clear is that requiring students to answer items on multiple choice tests or other types of assessments requiring product knowledge rather than process knowledge does not give us a clear view of what students are actually learning. Retrieval of facts (semantic memory) is primarily the responsibility of the left frontal lobe of the cerebral cortex, while recall of stories (episodic memory) takes place in the right frontal lobe (Zull, 2002). Recall of facts or memories does not indicate that students are capable of using this information in the process of creating new knowledge or making applications to problem-solving situations. It does not tell us that students are learning to understand their worlds or the people around them. High-stakes testing reveals only that students are able to respond to questions requiring little analysis as students choose the correct responses to superficial questions requiring minimal thought from an array of responses provided (Glasser, 1998; Zull, 2002).

To many, high-stakes testing may reveal that students have the "prerequisite knowledge" needed to build further information for particular subject areas. The fallacy in this thinking lies in the manner in which knowledge appears to be constructed and used. If students know that the earth is round but do not understand how this knowledge relates to ocean navigation, or how the curvature of the earth may affect one's perceptions of angles and planes when flying, then of what use is this knowledge? Is it possible that because students use the language of particular content areas teachers may view them as competent when in fact they have not mastered important concepts? These are questions that educators and policy makers must address without delay, before continued harm can come to students in the name of accountability.

Implications for Teacher Preparation Programs

Teachers form an important part of children's lives. They influence students in many ways, from understanding the world and learning how it operates, to understanding themselves and building their feelings of competence and positive self-esteem. How teachers act within the contexts of the school and community influences the moral development of their students (Costanzo, 2002). This is true for those teaching American Indian students as much, if not more so, since non-Native teachers represent the dominant culture, its values and mores, to American Indian communities.

Recently, my attention was brought to a situation in which a young teacher, sincere in her efforts to teach students about American Indians, told Native legends around a "campfire" in her classroom while dressed as an Indian Chief. She then used these same legends as materials to assist students with their reading fluency. This teacher was unaware that many Native Americans would perceive what she was doing as very offensive: (a) She portrayed American Indians stereotypically, and as a woman dressed as a man, which is not done in traditional communities; (b) she did not know about the spiritual aspect involved in becoming a chief; and (c) legends, the equivalent of stories in the Bible, Koran, or similar doctrines, are inappropriate to use for increasing students' fluency.

Native American children would have difficulty complying with this teacher's curriculum built on stereotypes that *Brown vs. the Board of Education* (1956) hoped to eliminate (Smith, 2004). The above example illustrates the importance of knowledge of culturally appropriate and relevant pedagogy for teachers serving all populations.

Teachers must be able to connect with their students and give them the support they need to flourish and grow in academic environments (Klug & Whitfield, 2003). This means they must respect their students and the communities in which students live, have high expectations for their learning, and understand the nature of the learning process and relevancy to students' lives and cultures. To prepare new teachers for schools serving American Indian students, teachers must understand what is considered to be appropriate and how knowledge is demonstrated in the community. Many talented teachers have left their schools because they were unable to make the necessary connections with the students and communities in which they taught (Zimpher & Ashburn, 1993).

When teachers do form bonds among school, home, neighborhood, and community, they foster a sense of ownership in schools and well-being for students (Maeroff, 1998). This sense of ownership of the educational process for American Indian students has been tenuous, as the government has repeatedly reasserted its right to determine the curricula and assessment measures used in schools serving Native students (Lomawaima & McCarty, 2002). NCLB (2001) and its attendant policies once again reinforce the idea that what serves the dominant culture best serves all citizens.

Within teacher preparation programs, decisions about who to admit and license must include consideration of candidates' dispositions to care for all students within the school setting. They must understand the importance of culturally relevant pedagogy, making connections with communities, creating emotionally safe environments in classrooms, as well as mastering the content for which they will be responsible. Only then will they be able to make good decisions in the future for their Native American students based on all the evidence and understandings of American Indian cultures.

References

Abedi, J. (2004). The No Child Left Behind Act and English language learners: Assessment and accountability. *Educational Researcher, 33*(1), 4–14.

Alaska Native Knowledge Network. (1998). *Alaska Standards for Culturally Responsive Schools*. Anchorage, AK: Author. Retrieved December 18, 2005, from www.ankn.uaf.edu/standards/standards.html.

Berliner, D. C. (2004). *If the underlying premise for No Child Left Behind is false, how can that act solve our problems?* Occasional Research Paper #6. Des Moines: The Iowa Academy of Education. Retrieved December 18, 2005, from www.shutupandteach.org/berliner.pdf.

Bernol, E. M. (2002). Three ways to achieve a more equitable representation of culturally and linguistically different students in GT programs. *Roeper Review, 24*(2), 82–88.

Bronfenbrenner, U. (1979). *The ecology of human development*. Cambridge, MA: Harvard University Press.

Bureau of Indian Affairs. (1998). *Establishing American Indian Ancestry*. Retrieved November 16, 2001, from www.doi.gov/bia/ancestry/ancestry.html.

Costanzo, P. (2002). Social Exchange and the developing syntax of moral orientation. In W. G. Graziano & B. Laursen (Eds.), *Social Exchange in Development* (pp. 41–52). San Francisco: Jossey-Bass.

Dewey, J. (1902). *The school and society*. Chicago: University of Chicago Press.

Frazier, G. W. (1993). *Urban Indians: Drums from the cities*. Denver, CO: Arrowstar.

Glasser, W. (1998). *The quality school teacher* (Rev. ed). New York: HarperCollins.

Gonzales, A. A. (2001). Urban (Trans)Formations: Changes in the meaning and use of American Indian identity. In S. Lobo and K. Peters (Eds.), *American Indians and the urban experience* (pp. 169–185). New York: Altamira Press.

Indian Education Act of 1972, 20 U.S.C. § 3385 *et seq.* (U.S.C. 2000).

Indian Nations at Risk Task Force. (1992). *Indian Nations at Risk*. Washington, D.C.: U.S. Department of Education.

Indian Self-Determination and Education Assistance Act of 1975, 25 U.S.C. § 450f. (U.S.C. 2000).

Johansen, B. E. (2000). Education: The nightmare and the dream. *Native Peoples Magazine, XIII*(1), 10–20.

Kappler, C. J. (1904). *Indian affairs: Laws and treaties Vol. II*. Washington, D.C.: Government Printing Office.

Klug, B. J., & Whitfield, P. T. (2003). *Widening the circle: Culturally relevant pedagogy for American Indian children*. New York: RoutledgeFalmer.

Leap, W. L. (1993). *American Indian English*. Salt Lake City: University of Utah Press.

Lomawaima, K. T., & McCarty, T. L. (2002). When tribal sovereignty challenges democracy: American Indian education and the democratic ideal. *American Educational Research Journal, 39*(2), 279–305.

Macqueen, A. (2000). Four generations of abuse. *Native Peoples Magazine, XIII*(1), 20–30.

Maeroff, G. I. (1998). Altered destinies: Making life better for schoolchildren in need. *Phi Delta Kappan, 79*(6), 425–432.

Meriam, L. (1977). The effects of boarding schools on Indian Family Life: 1928. In S. Unger (Ed.), *Destruction of American Indian families*. United States: The Association on American Indian Affairs. Original work published 1928.

Morgan, H. (2004). *Real learning: A bridge to cognitive neuroscience*. Lanham, MD: Rowman & Littlefield.

Nabovkov, P. (1978). Introduction to "The Treaty Trail." In P. Nabovkov (Ed.), *Native American testimony: An anthology of Indian and white relations, first encounter to dispossession* (pp. 147–152). New York: Thomas Y. Crowell.

National Reading Panel. (2000). *Teaching children to read: An evidence-based assessment of the scientific research literature on reading and its implications for reading instruction.* Washington, D.C.: National Institute of Child Health and Human Development.

No Child Left Behind Act of 2001: Reauthorization of the Elementary and Secondary Education Act of 1965, 20 U.S.C. § 6301 et seq. (U.S.C. 2001).

Pewewardy, C. D. (1994). Culturally responsible pedagogy in action: An American Indian Magnet School. In E. R. Hollins, J. E. King, & W. C. Hayman (Eds.), *Teaching diverse populations: Formulating a knowledge base* (pp. 77–92). New York: State University of New York Press.

Public Information Office. (2004, June 15). American Indian and Alaska Native Summary File (AIANSF). U.S. Census Bureau. Retrieved December 21, 2005, from www.census.gov/Press-Release/www/AIANSF.html.

Sheffield, G. K. (1997). *The Arbitrary Indian: The Indian Arts and Crafts Act of 1990.* Norman: University of Oklahoma Press.

Skinner, L. (1992). Teaching through traditions: Incorporating Native languages and cultures into curricula. In P. Cahape & C. B. Howley (Eds.), *Indian Nations at risk: Listening to the people* (pp. 54–59). Summaries of Papers Commissioned by the Indian Nations At Risk Task Force of the U.S. Department of Education. (ERIC Document Reproduction Service No. ED339588).

Smith, G. P. (2004, February). *Desegregation and resegregation after Brown: Implications for multicultural education.* Paper presented at the Annual Meeting of the Association of Teacher Educators, Dallas, TX.

Snipp, C. M. (2000, May). *Some alternate approaches to the classification of American Indians and Alaska Natives.* Paper prepared for the Executive Order 13096 National American Indian and Alaska Native Education Research Agenda Conference, Albuquerque, New Mexico.

Spring, J. (2004). *Deculturalization and the struggle for equality* (4th ed.). New York: McGraw-Hill.

Strom, K. M. (1994–2000). *Virtual Library: American Indians: Index of Native American resources on the Internet.* Retrieved November 15, 2001, from www.hanksville.org/Naresources/indices/Naschools.html.

St. Germaine, R. (1995). *Drop-out rates among American Indian and Alaska Native students: Beyond cultural discontinuity.* ERIC Digest (ERIC Document Reproduction Service No. ED3888 492).

St. Germaine, R. D. (2000, May). *A chance to go full circle: Building on reforms to create effective learning.* Paper prepared for the National American Indian and Alaska Native Education Research Agenda Conference, Albuquerque, New Mexico.

Testerman, J. (1996). Holding at-risk students: The secret is one-on-one. *Phi Delta Kappan, 77*(5), 364–365.

Vygotsky, L. S. (1978). *Mind in society.* Cambridge, MA: Harvard University Press.

Wadsworth, B. J. (1996). *Piaget's constructivism theory of cognitive and affective development: Foundations of* (5th ed.). White Plains, NY: Longman.

Wax, M. (1971). *Indian Americans: Unity and diversity.* Englewood Cliffs, NJ: Prentice-Hall.
Zimpher, N. L., & Ashburn, E. A. (1993). Countering parochialism in teacher candidates. In M. E. Dilworth (Ed.), *Diversity in teacher education: New expectations* (pp. 40–62). San Francisco: Jossey-Bass.
Zull, J. E. (2002). *The art of changing the brain.* Sterling, VA: Stylus.

Chapter 7

Accountability Systems' Narrowing Effect on Curriculum in the United States

A REPORT WITHIN AN ELEMENTARY EDUCATION TEACHER CERTIFICATION PROGRAM

Jennifer L. Snow-Gerono
Boise State University

Cheryl A. Franklin
Boise State University

> Jennifer Snow-Gerono, Ph.D., is an assistant professor in Boise State University's College of Education and is active in its teacher education program. Her areas of research emphasis include practitioner inquiry, language and literacy education, and professional development for educators. She has recently published in *Teacher Education Quarterly* and *Teaching and Teacher Education*.
>
> Cheryl Franklin, Ph.D., is a professor in curriculum, instruction, and foundational studies at Boise State University and teaches courses in social studies education and graduate courses in curriculum and instruction. She received her doctorate in curriculum and instruction from the University of Virginia. Her research interests include technology integration, social studies education, and curriculum and instruction in K–12 settings.

ABSTRACT

> This chapter outlines findings from a research study focused on the influence of the United States' No Child Left Behind (NCLB) policy on teaching and teacher education environments in the Northern Rocky Mountain Region. It describes how teaching and learning are

changing through the eyes of preservice interns and their mentor teachers and includes particular findings focused on the narrowing and co-opting of curriculum. Implications of the study's findings are discussed and a call is issued for the careful consideration of standards and accountability structures connected to national policy. The authors conclude that maintaining a partnership and collaborative perspective on teaching and education is of the utmost importance in order to respond to NCLB implementation in informed and productive ways that lead toward success for all teachers and learners.

Introduction: Seeking the Absent Presence

Grumet (1988) as well as Norris and Sawyer (2004) refer to the "presence of an absence" or an "absent presence" in curriculum. Perhaps Grumet puts it best when she states:

> Present is the curriculum, the course of study, the current compliance, general education, computer literacy, master teachers, the liberal arts, reading readiness, time on task. Present is the window. Absent is the ground from which these figures are drawn, negation and aspiration. Absent is the laugh that rises from the belly, the whimper, and the song. Suppressed is the body count, Auschwitz, Bhopal; even the survivors, the hibakusha of Nagasaki and Hiroshima, are invisible. Absent is the darkness and the light. (Grumet, 1988, p. xiii)

Grumet, of course, is referring to what is the stated curriculum for schools. But, she is also referring to a hidden curriculum. She is referring to what is being left out of the experience as much as she is pushing readers to consider the whys and hows of what is left unsaid. It is within this framework we propose to share our study of curriculum within the current context of the U.S. No Child Left Behind (NCLB) policy (U.S. Department of Education, 2002) and standardized testing movement that pervades at least one area in the Northern Rocky Mountain Region.

Teaching and teacher education potentially lack input into the implementation of policy and important decisions within public school contexts. When policy-makers and public school administrators determine changes at the classroom level without including voices of educators from that space, another present absence is created. Liston and Zeichner (1991) outline the importance of the "social conditions of schooling" to teacher education. If schools are to be places where young children and young adults learn tenets of democracy and participation within U.S. society, then teachers must be able to focus on the cultivation of agency and advocacy among their students, not to mention themselves. The presence of a curriculum that prepares students not only to perform well on standardized tests in reading and mathematics but also to engage in social policy and the construction and maintenance of smaller and more global communities is pivotal.

Education in the U.S. has been recognized as "at risk" since the 1980s (see, for example, A Nation at Risk, Carnegie Forum on Education and Economy, 1983). Currently, NCLB outlines standards and accountability mandates for public school districts, K–12 schools, students, teachers, and, in effect, teacher educators. The influence of NCLB on schooling in the United States is widely known to the general public due to news media and policy discussion. This targeted outreach to a dominant population that attends to media and policy discussions has been met with general and, sometimes, nostalgic acceptance, because emphasizing the "3 Rs" of reading, writing, and arithmetic reminds this population of their own educational experiences. However, these 3 Rs present an absence of the social studies and scientific study, which are equally necessary for well-informed and active democratic participation.

It is imperative that students are prepared as well-rounded individuals who integrate subject, self, and social in learning and becoming (Dewey, 1915). Indeed, a tension among standards, curricular areas and emphases, and teachers' beliefs may be strong when teachers and/or teacher educators feel they know what is best for children while at the same time feel they have their hands metaphorically tied due to their responsibility to prepare students for successful standardized test performance. The tensions between what teachers espouse they would like to include in curriculum and what they feel they can in their "new" teaching environment are explored more fully in the data excerpts that follow.

In this chapter, we will describe our situated context as teacher educators and curriculum workers and our research study focusing on school-university partnerships. This research explores how teaching and learning is changing through the eyes of preservice interns and their mentor teachers, and our particular findings focused on the narrowing and co-opting of curriculum. We will discuss implications of the findings that are geared toward curriculum in K–12 schooling and teacher education and issue a call for the careful consideration of implementation concerning standards and accountability connected to national policy. Teacher educators can no longer turn a blind eye to what happens in the very schools in which they share space in the initial preparation of teachers. We must take our roles as teacher educators seriously and emulate an inquiry stance toward education across the professional life span (Cochran-Smith & Lytle, 2001). If we value partnerships in teacher education, we must also value the more difficult negotiations of meaning and a present absence within curriculum and teacher education.

Framework for Study and Methodology

Numerous studies have examined the effects of state-mandated testing programs (especially those with "high-stakes" attached) on schools, teachers, and students. Research on these effects in various states yields both positive and negative results for teaching, learning, and curriculum (Abrams, Pedulla, & Madaus, 2003). This should not be surprising considering the variations in state testing programs, actual tests, and accountability connected to testing.

Madaus (1988) warns that testing may become the "end of instruction" rather than a "tool of instruction" when high-stakes tests, which connect teacher performance and student achievement, weigh too heavily on curriculum and pedagogy. Heeding multiple warnings of the dangers of standardized testing as a means of accountability in education (see, for example, Chudowsky & Pelegrino, 2003; Yen & Henderson, 2002; Miller, 2001; Holloway, 2001; Kohn, 1999), we, as two teacher educators, were determined to explore the influences of the initial implementation of NCLB legislation in a Northern Rocky Mountain region.

RESEARCH OBJECTIVES

The purpose of this research study was to describe perceptions of mentor teachers in elementary schools who work with preservice teachers from a local university as well as the perceptions of their teaching interns. Mentor teachers and their interns were asked to respond to survey questions and participate in focus group interviews focusing on how their lives in elementary schools/classrooms have changed as a result of new standardized testing requirements. The following research objectives are addressed in this chapter:

- Consider the impact of the current high-stakes testing environment on preservice teachers.

- Examine how standardized testing affects curriculum and instruction in elementary classrooms.
- Describe how standardized testing and accountability affect the teaching job satisfaction of elementary mentor teachers who serve as field placement mentors for preservice elementary teachers.
- Reconsider how to best prepare new teachers within the realm of standardized testing and accountability in local partnership schools.

As teacher educators, we believe these objectives uncover information that is imperative to our work in the initial preparation of teachers as well as efforts toward professional development and teacher education across the professional life span.

METHODOLOGY

In order to meet these objectives, we engaged in a study of the perceptions of mentor teachers and their interns. The study included distributing surveys that elicited narrative comments and conducting focus group interviews with mentor teachers. Participants included 106 mentor teachers (survey response rate 86%) and 55 preservice teachers (survey response rate 93%). Most preservice teachers had more than one mentor teacher. Sixteen mentor teachers also participated in one of two focus group interviews to provide more detailed information.

Narrative comments were invited throughout the surveys, and nearly all respondents included comments in some form or another. These narrative comments were entered into a Word document for analysis and triangulation with focus group data. Focus group interviews were audio-taped and transcribed verbatim using electronic transcription software. Additionally, narrative comments were coded and analyzed for themes by memoing the data and through multiple readings (Patton, 2004). Focus group transcripts were read by at least three analysts multiple times for coding and theme identification. The data focused on multiple issues, but the major focus of this chapter is the impact that increased testing due to NCLB implementation has on elementary school curriculum, according to these teachers.

SCHOOL-UNIVERSITY PARTNERSHIP

In line with calls in the teacher education field toward bridging theory and practice (Darling-Hammond, 1994) and creating partnerships among school-based and university-based teacher educators (Goodlad, 1990; Holmes Group,

1990; Teitel, 2003), professors at this university worked with practicing teachers and administrators in neighboring school districts to redesign their initial teacher preparation program around field experiences at multiple levels for elementary education undergraduate students.

Preservice teachers spend three years in partnership schools. They begin with tutoring and one-on-one participation with students then move toward small group and whole class instruction during a final professional year internship. One of the field experiences focuses on diversity. Another is based in a special education environment. Preservice teachers take elementary curriculum and instruction courses, based in literacy, science, social studies, mathematics, and classroom learning environments, usually while in their "Professional Year." This Professional Year internship includes a semester where students spend two or three days a week in an elementary school while engaging in 12–15 credits of university-based courses and a "student teaching" semester where the students spend the entire semester in an elementary school with weekly seminars to guide reflective practice. The conceptual framework for this College of Education focuses on reflective practitioners who:

> think critically about pedagogy, subject matter, and the needs and backgrounds of all students [and] . . . are guided by our professional understanding of the importance of reflection and the process by which it occurs. In practice, reflection requires educators to continually test ideas and hypotheses and to judge the worth of activities by careful observation of consequences. (Elementary Education Field Guide, 2005, p. 3)

This field-based program has been revised slightly in recent years to meet needs of the undergraduate students and state requirements for teacher certification, but the basis of reflective practice has remained the same.

PARTICIPANTS

Participants included 106 mentor teachers who work with elementary preservice teachers in school-university partnerships in the region and 55 of those preservice teachers. All of the respondents worked in partnership schools, so named by university and school partners during an effort to increase participation in initial teacher preparation and development across institutional boundaries (Teitel, 2003). With an interest in the Professional Development School movement (Holmes Group, 1990), this university began partnering more strongly with public schools interested in initial teacher preparation. Each of the mentor teachers in this study worked within one of these partnership schools at varying levels of mentorship.

Mentor teachers who responded to the survey had an average of 19 years teaching experience and had served as mentor teachers for an average of six years. The majority of teachers worked in first through fifth grades with a few kindergarten and sixth grade teachers as well as some specialists—for example, reading teachers. Nearly 30% of the mentor teachers taught in Title 1 schools, with 13% teaching English Language Learners (ELL) and children with identified special needs. This percentage of Title 1 and ELL students exceeds the state average. The preservice teacher field placements were distributed in grades K–6.

Throughout the past four years, elementary teachers in the state have been required to administer, on average, seven different local, state, and national standardized tests per academic year. Each school in this study, at the very least, administered the following standardized tests. Students are assessed on a state reading indicator test, kindergarten through 3rd grades, three times a year and rated on "reading fluency." The state standardized achievement test is administered in every grade 2nd–6th twice yearly with a mandatory "passing" regulation in 10th grade in order to complete high school graduation requirements. This state also has requirements for the regular administration of a standardized Direct Writing Assessment in 5th, 7th, and 9th grades and a Direct Math Assessment in 4th, 6th, and 8th grades. In the elementary grades, a sampling of students in grades 4th–8th also takes the standardized National Assessment of Educational Progress exams in math, reading, and science. In addition, most districts require monthly, quarterly, or semester tests that correspond to reading and mathematics basal curriculum.

As the teacher preparation program at this university worked to redesign components of its elementary education program, we believed it was necessary to learn about the influences of testing not only on the mentor and preservice teachers but also on curriculum and instruction. This data informs teacher education faculty in program and course content redesign. Likewise, this data could inform ways in which teacher education and partnerships need to work together to create learning situations that we know to be best for the initial preparation of teachers and, most importantly, for elementary school students as well.

Absence: Job Satisfaction

A majority of data from the returned surveys focused on the increased stress and pressure that mentor teachers feel along with an increase in job dissatisfaction. Most of the teachers' dissatisfaction comments allude to not so much an innate dislike or distrust of tests as a dislike of the intense focus on "all things tested." For example, a teacher shares:

> Testing and data are very important parts of learning and teaching; however we work with real little human beings and not tally marks on a piece of paper. We have to get through the affective filter before a great deal of quality learning can take place. We need a lot more balance in education—life is not just a test score. (Survey respondent 006)

Yet, more teachers suggest their worries concern professional longevity in a job that includes so much pressure:

> After 13 years, I still love my job. I still believe it's an important one and that I make a difference. But, in the last few years, it doesn't seem as enjoyable.... I have a growing concern about the long-term effects of all this standardized testing. I am afraid we are taking the joy and wonder out of education. I feel we are so concerned about raising scores that we focus more on rote memorization and test strategies than hands-on learning. I also think that the [State Reading Indicator] will produce a generation of readers who can read quickly, fluently, and excellently but hate to do it because they weren't taught to love books. They were taught to read as quickly as they could. (Survey respondent 030)

As can be noted in the respondents' comments, several teachers question the connection between standardized achievement test scores and student learning. One teacher describes this tension in the following way:

> Testing is turning educators to play the 'numbers' game. We must raise those scores even when it does not raise the students' learning. Testing has shifted a school's emphasis from individual learning to making the AYP (Adequate Yearly Progress). As a teacher, I am faced with abandoning what I know is good for learning or leaving the profession I enjoy. It is not a great position to be in. (Survey respondent 045)

The following comments indicate the presence in some of the teachers' responses calling for more of a balance in education:

> Assessments and standardized testing has a definite role and is a tool for teachers to evaluate student learning. However, the total emphasis on test results has a negative effect on students and teachers. The test is only one snapshot of the student's performance. We need to continue to focus on the total child and the growth that child has made—not just one test score. Teachers will continue to leave the profession if the current conditions and unrealistic expectations remain. (Survey respondent 063)

It comes through loud and clear within these teacher comments that they are not radical "anti-testing" or "anti-assessment" proponents. Therefore, the question persists: How may teacher educators best support teaching professionals who embrace high standards for their students at the same time that they recognize limitations of a local standardized test?

Absence: Breadth and Depth of Curriculum

Other findings connected to curriculum were also more important for their felt absence than any increased presence. Significant curricular and instructional changes have occurred as a result of high-stakes testing and accountability in this area. Mentor teachers acknowledged that they were strongly encouraged to ensure test objectives were covered in their curriculum and instruction. They were also working in contexts where they felt increased pressure to adjust instructional plans based on students' most recent test scores.

Our study also found that teachers were spending more time on test preparation with their elementary-aged students. One mentor teacher shared the feeling that she was not using grade-level appropriate language with her students because they needed to understand certain vocabulary in order to do well on standardized assessments. This presence of test preparation curriculum in content and skills is not in itself alarming until we consider the presence of a resulting absence. What happens to the curriculum and instruction that may have been in place prior to this emphasis on test preparation? Should teacher educators emphasize the importance of test preparation in initial teacher preparation curriculum or assist all educators in curriculum and instruction and the integration of test preparation skills throughout the school day? How do teacher educators intervene in schools where they also need to be supportive and emphasize a partnership toward education?

A powerful theme culled from the mentor teachers' narrative comments included remarks about the changes in curriculum due to new testing requirements. The majority of the comments mentioned the reality of "teaching to the test," and several other comments spoke specifically to the narrowing of curriculum in schools/classrooms. Mentor teacher respondents indicated that there was an increase in drilling students and preparing them with test-taking skills as well as teaching only content that they knew would be covered on the test. One teacher shared this feeling by writing:

> I am saddened at how 'driving hard' for test results has caused other activities such as art to virtually disappear. I rarely have time to dis-

cuss the Weekly Reader or current events. I have cut way back on reading aloud to my class. Social Studies and Science have been lowered in priority. (Survey respondent 002)

Another teacher commented on the narrowing of curriculum:

> We feel so much pressure that we spend most of the day teaching reading and math. Standardized testing is taking the fun out of teaching and learning. We drill and drill. . . . I don't teach science or social studies in a hands-on way. We learn science and social studies through our reading. . . . (Survey respondent 013)

Yet another teacher used an efficiency metaphor to describe how classrooms were changing due to new testing requirements in elementary schools:

> I feel the "real fun" of learning and exploring life has come to an end from the moment a child steps into a kindergarten classroom. We are forced to train these little bodies into a line of "factory workers" where they must learn to be drilled and skilled. Unfortunately, we [teachers] are also in this line of "factory workers." (Survey respondent 074)

Intern respondents also noted the narrowing of curriculum due to testing requirements:

> With [various test] results, my 2nd-grade mentor teacher and the rest of the 2nd-grade teachers were asked by the principal to remove such non-tested material (Reading Rainbow and Art, for example) from the curriculum. Or at least, not have both weekly. Instead, the teachers are to focus more on "upping" the test scores. . . . In my Block III placement, there was such an emphasis on tested readings that many of the students were beginning to hate reading. Unfortunately, the testing emphasis does not take into account each individual's needs. . . . Too many kids will be left behind, because they won't be encouraged to develop their own interests . . . or learn critical thinking skills. They'll rely on rote memorization. (Intern survey respondent 005)

An underlying theme in all of these comments is that the fun of learning is gone, and school may become boring for students. Although the joy found in teaching and learning can undoubtedly be considered an important motivation for new teachers to pursue this profession and a compelling reason for elementary students to come to school and engage in learning activities, how may teacher educators, as partners, help emphasize the serious hard work and dedication involved in both teaching and learning? What areas of curriculum should be more emphasized in elementary education and how could all educators best envision curricular areas in their schools?

We noted an integration of job dissatisfaction and increased pressure and frustration in the lives of teachers through all of our research data. The narrowing of curriculum and emphasis on test content and skills was influencing the overall dissatisfaction that teachers described. Likewise, teachers' overall dissatisfaction and the increased stress and pressure they felt played into each other for a dynamic relationship that worked against creating environments conducive to teaching and learning. Although teachers who participated in this study allowed for the benefits of standardized testing and the importance of accountability in teaching and learning, they were hesitant to embrace the high-stress environment that emphasized a narrowed curriculum and skill-based teaching and learning. Perhaps one teacher best summed up the skepticism many teachers could be feeling across the United States:

> It is difficult to express your ideas, thoughts, and feelings. I think it's great to assess children, but there is a point of over assessing them. I've noticed an increase in behavior and defiance toward wanting to work. Where do we establish the blame? Teacher: pushes too hard? Tests: too much too often? Lifestyles: parents model it? Computers/TV/videogames? When is enough, enough for these youngsters? We will see something big happen with these kids in the next few years. Good or bad? Yet to be decided! (Respondent 019)

The narrowing of curriculum in these teachers' classrooms and students' lives leads to an emphasis on tested content and creates the presence of the absence of non-tested curricular areas. What happens when students do not spend adequate time on scientific method and discovery? Better yet, what happens to society when students do not study history, geography, global connections, civic participation, and democratic responsibility? Perhaps what we should be asking is what will happen when these content areas are taught via reading instruction—a practice that teachers indicated is becoming so tedious for students that they do not learn to enjoy the process of reading let alone the content. This present absence leads us to the absent presence of a dominant discourse that overtakes curriculum more thoroughly than any "back to basics" education has in the past.

Presence: Curriculum is Increasingly Co-opted in Elementary Classrooms

Another finding related to curriculum in this study was an increased presence of large-scale curriculum programs. For example, scripted programs are used to teach reading and mathematics. These programs are geared toward increasing standardized test performance of students. However, many literacy scholars argue for a balanced approach to literacy instruction (Fountas & Pinnell, 1996). Therefore, a dominant curriculum mandating un-differentiated instruction co-opts other forms of curriculum and instruction that many teacher educators espouse and/or model in their classrooms. Asking for and emphasizing constructivism (Richardson, 1997) and differentiated instruction (Tomlinson, 1999) in teacher education programs appears to be at odds with the whole-group instruction increasingly being implemented at the classroom level. Does that mean teacher education curricula is being co-opted by federal mandates and K–12 curriculum and policy implementation? If it does, what does that mean for teacher education? Or even more importantly in our case, what does that mean for teacher education partnerships among schools and universities?

Another presence of co-optation may come through large-scale curricular initiatives like character education and civic education. These programs are being implemented in local elementary schools and included as social studies on the daily schedule. We do not mean to take anything away from properly implemented character or civic education programs, but mandating their implementation and then calling them the social studies curriculum co-opts doing social studies in meaningful ways. It often trumps history, community-building, authentic inquiry and analysis, and geography. Such programs may create a dominant discourse for social studies curriculum experiences that marginalizes disciplined inquiry into history and the social sciences. Inquiry is generally absent from Weekly Reader activities or the discussion of the character trait for the week. Recall Mentor respondent 002 suggested the Weekly Reader was her social studies curriculum and even that was being relegated to "when there is time."

The interns also recognized the tensions within curriculum and teaching that their mentor teachers were feeling. Intern respondent 012 shared:

> My mentoring teacher said that she used to do enrichment activities and units of discovery every year. She no longer teaches anything like this. If it is not tested on a standardized test—there is not time in a classroom for "that" kind of learning.

Several of the preservice teachers did share that they were not anti-assessment. They appreciated testing for its ability to show growth in their students or as a means for informing them about important issues and concepts to revisit in their teaching. The problem was when the standardized testing appeared to co-opt student learning with understanding. One particularly despondent preservice teacher noted that she would not be going into teaching after graduation:

> The focus on standardized testing and the resulting pressure, along with the de-emphasis of higher thought processes among students, are the main reasons I am getting out of this industry. I do not feel an emphasis on standardized testing improves schools; I think it demands a focus on information regurgitation, especially when important subjects like social studies, art, and music are being removed from schools to accommodate the desires of our government. In addition, class time is being taken so that kids from a variety of backgrounds and experiences can be forced to be the same as each other. (Intern respondent 014)

This intern's comments suggest great disillusion with her public school classroom experience. Another intern who worked in a charter school, which focused on experiential learning, was quite pleased with her placement but apprehensive about entering the workforce as an educator. Intern respondent 007 stated:

> I have been fortunate to be placed at a school where test results are looked at as one indicator of a student's knowledge and/or learning. The teachers spend a great deal of time teaching students in authentic (real world) type situations and develop their higher level thinking skills. . . . I am very uneasy about getting a job in a school that pushes "teaching to the test" when I can see that there are other ways that work just as effectively to get test results and develop a student's deeper understanding.

How can teacher education best tease out the tensions this intern is experiencing so that novice teachers come to understand good teaching and learning practice while at the same time navigate the realities of their classroom, school, and district demands?

Intern teachers indicated that they taught social studies or science approximately one to two times per week in their placement classrooms. This occurred because these disciplines are not tested; therefore, mentor teachers may have felt more at ease in allocating only 60–90 minutes a week to social studies and science. A decrease in time spent on certain areas of curriculum silently states that other areas of curriculum are more important. Rather than calling for standardized tests in every content area, which would be a plausible argument in this case, we instead choose to analyze how the presence of the absence of social studies and (for now) science curriculum on standardized tests co-opts the importance of their meaning and worth in the minds of elementary students, their families, and perhaps even their teachers. If we hang on to the necessity of these curricular areas in teacher education, what should our role be in voicing our concerns over the present absence? How do we also maintain concern for music and art and the experience of school, the hidden curriculum, in the midst of fighting against a dominance of some skills and knowledge at the expense of others?

Implications for Teacher Education and Curriculum Workers at All Levels

Coble and Azordegan (2004) suggest seven strategies for teacher educators in living within NCLB implementation. They advocate for collaboration with public schools to "create high-quality alternative licensure programs" (p. 8) while at the same time arguing for school-university partnerships, the inclusion of arts and science faculty in initial teacher preparation, increased collaboration with community colleges, professional development for teachers, and the need to stay informed and to have rigorous admittance policies for teacher education students.

As we work within varied levels of school-university partnership structures, we believe it is important to maintain this presence in elementary classrooms so that, in offering to provide professional development opportunities, we may address not only the need for increased assessment literacy among all educators but also the need for noting the present absence of curricular areas within standardized testing implementation structures. Likewise, if we maintain a partnership and collaborative perspective on our work with/in elementary classrooms, we may also have a better chance of being heard as a "critical friend" rather than a "naysayer" or obstructionist when wondering about NCLB implementation or its trickle down effects in elementary classrooms.

One reason we conducted this research study was to inform our work with preservice teachers. We believe in strengthening partnerships between university and school-based education so that novice teachers are prepared to work with and within the environment of high-stakes testing.

We examined how standardized testing affects curriculum and instruction in elementary classrooms. What we hope to make clear is that the presence of a one size fits all curriculum connected to standardized testing creates a narrowing of the curriculum and a present absence. We need to work with educators at all levels in order to carefully consider standards and curriculum implementation for learning. Likewise we need to consider the increased job dissatisfaction of mentor teachers in elementary schools. How can we work together to increase job satisfaction and the agency of new teachers to change work environments in order to increase student learning and attitudes toward teaching?

Information from this research study causes us to reconsider how to best prepare new teachers within the realm of standardized testing and accountability in local partnership schools. Teacher education must be about preparing thoughtful, inquiring educators who consider meaningful learning opportunities for themselves and their students. This is where energies need to be focused. If teacher education programs feel vulnerable, they could become more productive by reaching out for authentic partnerships with public school classrooms rather than retreating into denial or engaging in combat with other entities. Grumet (1988) states:

> Underneath every curriculum, which expresses the relation of the knower and the known as it is realized within a specific social and historical moment, is an epistemological assumption concerning the relation of the subject and object. In an attempt to understand how we come to have and share a world, the various epistemologies relegate differing weights to consciousness and facticity. Each epistemology offers a negotiated peace between these two competing terms to account for this intersubjective construct, this ground of all our cognitions, "this world." (p. 15)

NCLB implementation creates a space where teacher educators and curriculum workers may come together to uncover the presence of "facticity" as well as consciousness and meaning so that peace may be negotiated and meaningful learning at all levels may occur.

References

Abrams, L. M., Pedulla, J. J., & Madaus, G. F. (2003). Views from the classroom: Teachers' opinions of statewide testing programs. *Theory into Practice, 42*(1), 18–30.

Carnegie Forum on Education and the Economy (1983). *A nation at risk*. Washington, DC: The Forum.

Chudowsky, N., & Pellegrino, J. W. (2003). Large-scale assessments that support learning: What will it take? *Theory into Practice, 42*(1), 75–84.

Coble, C. R., & Azordegan, J. M. (2004). The challenges and opportunities of the No Child Left Behind act: Seven strategies for teacher educators. *Action in Teacher Education, 26*(2), 2–14.

Cochran-Smith, M., & Lytle, S. L. (2001). Beyond certainty: Taking an inquiry stance on practice. In A. Leiberman & L. Miller (Eds.), *Teachers caught in the action: Professional development that matters* (pp. 45–58). New York: Teachers College Press.

Dewey, J. (1915). *The school and society*. Chicago: The University of Chicago Press.

Fountas, I. G., & Pinnell, G. S. (1996). *Guided reading: Good first teaching for all children*. Portsmouth, NH: Heinemann.

Grumet, M. (1988). *Bitter milk*. Amherst: The University of Massachusetts Press.

Hargrove, T., Walker, B. L., Huber, R. A., Corrigan, S. Z., & Moore, C. (2004). No teacher left behind: Supporting teachers as they implement standards-based reform in a test-based education environment. *Education, 124*(3), 567–572.

Holloway, J. H. (2001). The use and misuse of standardized tests. *Educational Leadership, 59*(1), 77–78.

Holmes Group (1990). *Tomorrow's schools: Principles for the design of professional development schools.* East Lansing, MI: Holmes Group.

Kohn, A. (1999). *The schools our children deserve.* Boston, MA: Houghton Mifflin.

Liston, D. P., & Zeichner, K. M. (1991). *Teacher education and the social conditions of schooling.* New York: Routledge.

Madaus, G. F. (1988). The distortion of teaching and testing: High-stakes testing and instruction. *Peabody Journal of Education, 65*(3), 29–46.

Miller, D. W. (2001). Scholars say high-stakes tests deserve a failing grade. *Chronicle of Higher Education, 47*(25), 14–16.

Norris, J., & Sawyer, R. (2004). Hidden and null curriculums of sexual orientation. In L. Coia, N. J. Brooks, S. J. Mayer, P. Pritchard, E. Heilman, M. L. Birch, & A. Mountain (Eds.), *Democratic responses in an era of standardization* (pp. 139–159). Troy, NY: Educator's International Press, Inc.

Office of Teacher Education. (2005) *Elementary education field guide: Learning for tomorrow teaching for today.* College of Education, Boise State University.

Patton, M. (2004). *Qualitative evaluation and research methods.* Newbury Park, CA: Sage Publications.

Richardson, V. (Ed.) (1997). *Constructivist teacher education: Building a world of new understandings.* London: The Falmer Press.

Teitel, L. (2003). *The Professional Development Schools handbook: Starting, sustaining, and assessing partnerships that improve student learning.* Thousand Oaks, CA: Corwin Press.

Tomlinson, C. (1999). *The differentiated classroom: Responding to the needs of all learners.* Alexandria, VA: Association for Supervision and Curriculum Development.

U.S. Department of Education. (2002). No child left behind act of 2001. Retrieved May 29, 2002, from www.ed.gov/legislation/ESEA02/.

Yen, W. M., & Henderson, D. L. (2002). Professional standards related to using large-scale state assessments in decisions for individual students. *Measurement & Evaluation in Counseling & Development, 35*(3), 132–143.

CHAPTER 8

The Impact of Standardized Testing on Teachers' Pedagogy in Three 2nd-Grade Classrooms of Varied Socio-Economic Settings

Annapurna Ganesh
Arizona State University

Annapurna Ganesh, Ph.D., is a faculty member of the Mary Lou Fulton College of Education at Arizona State University. Her areas of research interest are early childhood assessment, teacher pedagogy and teacher education, family-school partnerships, use of technology in early childhood, and the educational needs of children with diverse abilities.

ABSTRACT

In this chapter I attempt to make sense of the impact of standardized testing on teachers' pedagogy. I used interviews and day-long classroom observations in three second-grade classrooms located in a variety of socio-economic neighborhoods in one school district to examine the impact of standardized testing on teacher's pedagogy and to explore whether the socio-economic setting of the schools make a difference in the impact of the standardized tests.

Introduction

"It has changed us [teachers], because that [test scores] is all anybody looks at. . . ." This statement, culled from a longer passage of text, illustrates Mrs.

Mole's (one of my teacher participants) response concerning how testing has impacted her pedagogy and how the public views accountability and teachers. In this chapter, I examine the influence of achievement testing on teachers' pedagogy in early childhood classrooms where foundational skills are still being developed. I explore Mrs. Mole's pedagogy as well as the pedagogy of two other teachers in second-grade classrooms situated in different socio-economic neighborhoods. I have chosen this topic because standardized testing in early childhood education and its consequences intrigue me as an educator. As an early childhood educator, I wrestle with several questions every day. I wonder whether the items on the standardized tests are clear to young children. I ponder whether they address the particular educational concerns of teachers of young children or of parents. I consider whether the tests as a whole provide useful information about individual children and/or the whole class. I question if they help children in their learning and support this plethora of children's intentions as learners. I struggle with essential information provided to children's parents. These queries led me to study the impact of standardized testing on teacher pedagogy.

Testing Backdrop

Charles Darwin, Lawrence Frank, and G. Stanley Hall were leaders in the development of the child study movement that emerged at the turn of the 20th century (Wortham, 2005, p.4). The study and measurement of young children today has evolved from the child study movement to the development of standardized tests we use today. In the early 1900s, when institutions of higher education began seeking students from different parts of the nation, they faced a challenge with the evaluation of high school transcripts. To allow fairer comparisons of applicants, the Scholastic Aptitude Test (SAT) was established (Cronbach, 1990). Gradually, as public school education began to expand, there developed a need to determine the level of instruction and pace of instruction provided. To make these judgments, objective tests were developed without regard to the socio-cultural aspects of the student population (Gardner, 1961). The objective tests grew out of the need to sort, select, or otherwise make critical decisions about children and adults (Wortham, 2001).

Based on the work of Binet and Simon, American psychologists began to develop specific intelligence tests. Educators in particular welcomed the opportunity to use precise measurements to evaluate learning. It did not take a long time for standardized tests to be designed to measure school achievement. The after effects of World War II brought about a steady growth in the development of centralized testing. Statewide tests were administered in schools, and tests were increasingly used at the national level (Wortham, 2005, p.6). Expansion in the use of educational testing led to the establishment of major corporations that could assemble the resources required to develop, publish, score, and report the results of testing to their ever growing clientele.

A new education reform movement began in the 1980s in the aftermath of *A Nation at Risk* (Meisels, Steele, & Quinn-Leering, 1993). It was accompanied by an increased emphasis on testing. Efforts to improve education at all levels included the use of standardized tests to account for what students were learning. At that time, minimum competency tests, achievement tests, and screening instruments were used to ensure that students from preschool through college reached the desired educational goals and achieved the minimum standards of education that were established locally or by the state education agency (Wortham, 2005).

Throughout their schooling years, children are administered many different kinds of tests. Some tests determine their grades for each reporting period; others are achievement tests, IQ tests, or tests for admission to an educational institution. Standardized tests have changed the field of education in recent years. In a standardized test a set of predetermined questions are given to large numbers of students under the same conditions, such as time limit, and scored in the same way (Swope & Miner, 2000). They usually refer to multiple-choice tests given statewide or nationwide and tend to be "norm referenced," with the main purpose to compare and rank students (Swope & Miner, 2000, p.140). Norm referenced tests are designed to compare a student's score against the scores of a sample group called the "norming group." Most norm referenced tests are multiple-choice, although some include short-answer questions. One or two wrong answers can drastically change a score (Swope & Miner, 2000, p.140).

In this research study, the impact of standardized testing on teacher pedagogy in three second grade classrooms from the same school district in Arizona was investigated. Literature reveals that the results of standardized tests show that children attending schools with a high percentage of student population availing of the free and reduced price lunch have lower performance scores than children attending schools with a low percentage of free and reduced price lunch. Kohn (2000), for example, observed that, "the main thing they [the test results] tell us is how big the students' houses are" (p.7). Research has repeatedly found that the amount of poverty in the communities where schools are located (along with other variables having nothing to do with what happens in classrooms) accounts for the great majority of the difference in test scores from one area to the next. As mentioned earlier, the intent of this study was to understand the impact of standardized testing on teacher's pedagogy and to consider whether socio-economic settings of schools make a difference on the impact of standardized tests.

Methods and Data Sources

Using the statistics from the State Department of Education, School Nutrition Program, and Percentage of Free and Reduced Lunch Report, I identified three schools serving student populations from varied socio-economic backgrounds in one school district. I spent considerable time during the school year doing participant observations in a second grade classroom at each of these schools. In order to maintain anonymity I have used pseudonyms for my participants in this chapter.

In the particular school district in question, the following assessments were required at the second grade level:

- Stanford 9 (standardized test) (reading, vocabulary, and two portions of math; assessment in the spring)
- District writing (six trait writing test assessed in the fall and in the spring); District math assessments (six math standards assessed over the school year)
- Running records (assessment of reading and reading comprehension, assessed over the school year)
- High-frequency words (vocabulary words assessed over the school year)

I interviewed the three teachers extensively to gain information about their background, teaching experience, philosophy of teaching, views about standardized testing, and the impact of standardized testing on their pedagogy. I also gathered artifacts from the school and classroom, including newsletters and samples of student work. I furthermore audio-taped the interviews to assist me in capturing the substance of the conversations. To better understand the testing experiences of students and teachers, I requested teachers to share stories with me. Stories bring out the reality around testing to which people in the world outside the classroom are oblivious. The stories that were shared with me gave me a better perspective on the lives of the students, and unfortunately, the whole child is not taken into consideration when they are tested on standardized tests. I also looked at the performance of the three schools from the varied socio-economic areas on the Stanford-9 achievement test (SAT-9) given in the spring semester of the school year. I gathered from the stories related by my participants and from my experiences as a classroom teacher that school life is intertwined with family and community life and does not stand separate. The following is an excerpt from one of the stories shared with me during my study:

> He was by far my brightest student; he was just an amazing child. He had won all kinds of science awards with his project that he had submitted to NASA (National Aeronautics and Space Administra-

tion) and different places. He scored extremely low on his SAT-9 that year and the reason for that was his house had been robbed the night before. My student had informed me of his situation before he took the test; I requested that he be allowed to abstain from taking the test, because of his experience and that he needed time. I was told, "No, he is here, (and) he has to take the test." Based on his scores they (the school administrators) didn't want to accept him in the gifted program during the following school year, because his scores were low.... It felt like he was being cut out based on his performance that day; and that is just a crime. A whole year's worth of learning is not one day. We all have our bad days.

In the aforementioned passage, one of my teacher participants narrated one student's testing experience and the reality of how it affected his placement—which appears antithetical to developmentally appropriate practices. In the position statement on developmentally appropriate practices (DAP), the National Association for Education of the Young Child (1996) declared that

> developmentally appropriate practice requires that teachers of young children integrate the many dimensions of their knowledge base. They must know about child development and the implications of this knowledge for how to teach the content of the curriculum—what to teach and when—how to assess what children have learned, and how to adapt curriculum and instruction to children's individual strengths, needs, and interests. Further, they must know the particular children they teach and their families and be knowledgeable as well about the social and cultural context. (p.10)

The guidelines for DAP address five interrelated dimensions of early childhood professional practice: creating a caring community of learners, teaching to enhance development and learning, constructing appropriate curriculum, assessing children's development and learning, and establishing reciprocal relationships with families. Through my study I found that early childhood educators in heterogeneous settings entertain different notions of developmentally appropriate practices. I observed several practices, including play, direct instruction, learning centers, and team teaching, to name a few.

Mrs. Mole's Second-Grade Classroom

Mrs. Mole taught at a school where 15% of the student population was enrolled in the free and reduced lunch program. She had a bachelor's degree and lots of extra course work not tied to a particular degree, among which there was significant gifted education coursework. Mrs. Mole has 13 years of teaching experience, 5 of those years teaching second grade. In addition to teaching gifted children, she has taught in multi-age classrooms, bilingual classrooms, and in traditional classroom settings. Mrs. Mole describes her teaching philosophy in her own words below:

> It is pretty simple—to help students with whatever they need, to help them learn whatever way they learn best, and to be the kind of person to do whatever they need to learn. And I get to assess that of course; they don't get to tell me what they need to learn. I could say, "You look like you need more spelling or you look like you need more help with this." So, I have to be the judge of that. I would like to meet every child's needs, whatever they need individually.

In Mrs. Mole's classroom, the students worked independently for the most part. Mrs. Mole wrote the daily schedule on the white board, which guided them and also helped them stay on track. She mingled with the students and worked with individuals if they needed assistance. Most of the activities in the class took place in small groups; large group instruction and discussions were a smaller part of the routine. Mrs. Mole mentioned that she had strong parental support in her classroom. In her room of 24 students, she had 3 students who were on the free or reduced lunch program.

When asked about assessment techniques in her classroom, Mrs. Mole discussed them in the following way:

> I do a lot of written observations on students. I make notes in my monitoring notebook about students, what has happened through the day. I give quizzes, written quizzes; sometimes I'll give them group quizzes which they enjoy, and I can quickly assess if they are getting it [concepts] or not. Like odd and even today, I could see they did not really get it and that I probably went too fast when I was teaching the concept. So, I will readjust and tomorrow we will start back and go a little slow on it. I use a lot of observation and discussion in our circles and in our journals. I get most of my information that way.

When I observed Mrs. Mole's class, I noticed her making quick notes to herself during class time. She mentioned to me that she constantly jots notes during the day to help her keep track of student growth and areas that need improvement. She writes the student's name on the note and at the end of the day places the entries in the student's individual folder that she maintains and updates with information regularly. One day, for example, there was an incident that took place in the classroom between two boys. She had the boys discuss the incident with her and also mentioned to me that she would include the proceedings of the incident in her reflections in her monitoring notebook.

Journal writing in Mrs. Mole's classroom took place between the teacher and the students. Mrs. Mole also had a journal of her own in which she made entries during journal time. Each day a student got the opportunity to respond to the teacher's journal entry. I asked Mrs. Mole about the kinds of assessments she would like to use in an ideal situation. Her response is outlined in the following passage:

> Ideally, I would like assessments to give students the opportunity to verbalize what they are thinking or to sketch what they are thinking. Any way other than a forced one way would be good. There are some [students] in this classroom who will totally enjoy bubbling in the bubbles . . . that will be their thrill of the year, taking the Stanford-9. But, others are not going to show what they really know. And I think that it is so unfair, that I have to gear down my teaching style come January and I have to practice with them in that format, so they will feel comfortable. I am doing it for them, I don't care what they score, and I just want them to be comfortable doing it in that fashion, because I don't ever ask them to show me their knowledge by bubbling in!

Mrs. Mole, it seems, did not incorporate activities involving "bubble in" exercises as part of the students' regular curriculum. What she did do, in her words, was to

> ask them to explain it in a complete sentence or ask them to sketch it to me, or act it out with a partner or to even just talk to the partner and have them explain it to me, or even musically. There are all these different ways that they can express themselves and they become used to expressing themselves throughout the year, and that is happening in every classroom in our school. It is becoming a habit, to have choice and to have freedom, and so then to limit your freedom, and show your knowledge in only one way, it is just horrible. I do want them to be assessed, I do feel that that is important, I don't know how you can get out of being standardized, but I wish that we could. I wish that schools could just be trusted in reporting what their students know.

Here, we can see that Mrs. Mole was not anti-assessment or anti-accountability. Rather, the concerns she raised had to do with freedom and trust.

In response to asking Mrs. Mole how standardized testing had impacted her pedagogy, she addressed this question both from her point of view and from that of her school. She focused mainly on the interpretation of the results, especially the way teachers are asked to interpret them by the administrators:

> It has changed us [teachers], because that [test score] is all anybody looks at. The parent's everyday comment about, "Oh, these are wonderful projects." But the bottom line is they are going to pick a teacher who has higher scores. I totally disagree, because no matter who you are it is the students you have and they are going to do as well as they can. I think we can teach them and help them grow. I don't think they are going to be strong in all areas, at all times, in all grade levels, based on the teaching. It is unfortunate, I think that they [the scores] are misinterpreted; I think they [the results] are used way too much. I think people don't even look at the results appropriately.

She then went on to explain how she, as a teacher, would use quantitative results for decision-making: "We need to look at how a particular student grew from one year to another. If reading comprehension was that child's strength last year, was it still strong in third grade? Were they still strong in reading comprehension in fourth grade?"

However, as she explained, data from the test scores were being used differently in her school:

> When we look at the scores as a whole, what really happens is, I am told, "You weren't very strong in patterns, so you need to really hit patterns this year." The fact is that I don't have those students this year; those students are now in third grade. The kids that I have now may already be strong in patterns. They do this every single year.

She explained how the administrators made the teachers look at how their class performed and how their students did and find out their own instructional strengths and weaknesses from the scores: "But, do they [the administrators] realize that now those students have moved on, the present group is not the one that did not get it. There are lots of other ways the scores are misinterpreted. But, that is just one way."

Once again, Mrs. Mole returned to her musings, this time weighing the advantages and disadvantages of standardized tests. She then continued:

> Maybe the scores can be helpful in some way, but the way they present it and tell you to improve is not how we should be looking at it. I look at the student scores and now I focus on individual needs. I don't think people look at the scores correctly and they do not have any reason to judge and any reason to place our salary based on scores. We cannot be judged by the performance of children who got robbed the night before.

In the end result, it appears that another of Mrs. Mole's major concerns is how one individual—a teacher—is judged by the performances of other people—

students—who themselves are affected by factors in their environment that are out of their control.

Mrs. Riser's Second-Grade Classroom

Mrs. Riser's school had 33% of its student population participating in the free or reduced price lunch program. Mrs. Riser has a master's degree in education and her initial teaching certification. Her teaching experience spans over 14 years, and she has taught second grade for 6 years. Mrs. Riser's teaching philosophy is outlined below, in her own words:

> Basically, someone told me something a long time ago, or I read it somewhere, that children will never remember everything you taught them. And I saw that with my third and fourth grade multi-age class. . . . I would look at them and I would review something as simple as an adjective. And they would look at me like they didn't know what I was talking about. I could actually remember the month we studied adjectives and what the project was. They'll never remember everything you teach them, but they will remember how you treated them. Teaching is getting to know your students. Part of knowing their story is not to judge them or their parents, because I need that rapport with them [the parents] to work together as a team.

Mrs. Riser's class followed a structured routine. The students seemed familiar with the structure and knew what they were going to be doing during their day. Mrs. Riser used the basal reader provided for her class, because she found that the basal provided for both the enrichment and remediation of her students. Mrs. Riser mentioned that she had fairly strong parental involvement that year. In her class of 22 students, she had 10 students on free or reduced lunch. Among those students, there were two students who were of limited English proficiency.

Mrs. Riser described the assessment techniques she uses in her present day classroom in this way:

> I still do spelling tests and written math tests to show parents the growth of their children, and a reading test every nine weeks. And I don't do anything like a pop quiz, where I put them on the spot and make them terribly nervous; I wouldn't do that. I like to show films to tie in with my curriculum. I tell them to get out a page and take notes about the film or write some words to help them remember what they want to write about in their journals. [Mrs. Riser mentioned that she had already shown her students note-taking techniques.] So, journal writing is another way I assess my students' understanding of concepts.

When I observed in Mrs. Riser's class, I had opportunities to see her students work in groups on various activities. Mrs. Riser walked around and mingled with all the groups and observed individual participation. They also did worksheets individually and as a whole group. In an ideal setting, Mrs. Riser would prefer to use the following assessment techniques, as she explains below:

> For second grade, I would like to see if maybe someone could help me do observations with my students. Help doing that, because that does take time; you know I wouldn't mind an aide coming in here and working with a group, telling me where [the level] they are at. So then tomorrow when I want to assess them I know where they are at, to have some help in doing that, because it does take a lot of time. And it's such a visual situation that I wouldn't want to assume, because they were doing something that day, and didn't understand it, that they didn't totally understand the concept. So, I would like maybe some help in getting to that point.

Mrs. Riser addressed the question concerning how standardized testing had impacted her teaching pedagogy from the point of view of test preparation and how preparation for the test diminishes children's joy of learning:

> Well, we often do something here that I don't understand why we do it, and I'm sure that the children benefit from. Before we give the SAT-9 we practice a little using a test preparation guide. It's test prep; a lot of second graders have never filled in bubbles, in taking a test. I don't use it as a timed test. We are supposed to time it, but I figure they'll have enough of that in April. So, I use it as a practice. . . .

She then continued her reservations by describing the pressure she feels that schools are under:

> But when you are doing that, it takes away from other things you would have planned. It takes a good hour a day to do a section. You know with the influx of charter schools, parents want to know if the scores are high; they look at them when they buy in the neighborhood. If we are going to do that then we have to do this test prep.

Mrs. Riser then returned to "the joy of learning":

> And it takes away time from fun activities that would give them the joy of learning. That is how it has affected me. I get very frustrated. I understand why I am doing it. I can see it benefits some children, not all of them because I have seen the brightest children on the day of the test after we spend a whole month preparing they just start guessing.

She provided an example of a student's experience to emphasize her point about the time involved in test preparation:

> I am really good friends with a first grade teacher, and she and I think alike. She came to me and said that a student broke [out] in tears; this was on a test prep day, because she didn't understand where she was at, what she was supposed to be doing. She's got 26 first graders. It just sends a message to them [the students]; and here they are in school feeling comfortable, and all of a sudden they don't know something, they begin to feel insecure.

Mrs. Riser concluded her response to the question, with a statement to summarize the impact of testing on her pedagogy: "That's how it has affected me, more time on preparation for the test, and building a feeling of insecurity." In this statement, Mrs. Riser has captured her experience of accountability testing, which included some affordances but also, in her view, constraints that were not necessarily developmentally appropriate for emergent learners in early childhood classrooms.

Mrs. Rosen's Second-Grade Classroom

Mrs. Rosen taught at a school that had 86% of the student population enrolled in a free or reduced price lunch program the year I conducted my study. Mrs. Rosen has a master's degree and multiple hours of coursework on top of that. She was working on her English as a Second Language endorsement. Mrs. Rosen has a reading endorsement and a special education certificate. She has 16 years of teaching experience and has taught second grade for 8 years. She has experienced teaching special education for a few years; in fact, she was a self-contained teacher of learning disabled students. She also has played a role in the district office as a resource teacher and has had the opportunity of teaching multi-age classrooms. For the past two years or so, she has taught a straight second grade class.

Mrs. Rosen responded to my query about her philosophy of teaching with the following:

> I think that children learn developmentally and that they don't all learn at the same rate and that you need to go with their speed. I think learning needs to connect with the real world, rather their real world. I think it needs to be brain based. I think they need to feel safe and they need consistency. As crazy as their lives may be when they come to my room, I want them to feel comfortable. So it needs to feel like a home. It's not necessarily a democracy at all, or a benev-

> olent dictatorship, where the mom is in charge. [Laughs.] But, they feel like they can voice, and we discuss and problem solve together as a family does. It's kind of a family. They need to feel like they are a part of their learning, so they need to take some ownership and help decide where we are going. They need to have some part in deciding where the curriculum goes or how the topics flow. I really believe that reading is going to be the basis of a lot of things and language development is the key. So, we spend a lot of time on that. That is basically my philosophy.

From my conversations with Mrs. Rosen, I found that she had a wide range of abilities in her group of students and preferred to conduct small group instruction as opposed to whole group instruction. She worked with small groups of students, providing them instruction at their particular levels of ability. Her class had 27 students of whom 11 were English as second language learners. From my observations, it seemed like she had more than a handful of students who were limited English proficiency students. Of the 27 students in her class, all but one was on free or reduced lunch. Mrs. Rosen described her assessment techniques in the following manner:

> We use all of these (referring to the required assessments on second graders, as discussed earlier in this chapter), and I also use math tests—mini tests at the end of units. I do spelling tests once a week. I also do like a practice writing sample, what we call power writing on Fridays. I look at them and see how they are doing. And I constantly do observation, and take mental notes. The other thing I do is AR (Accelerated Reading) using the computer. They read the books, take the test, and the results are graphed on the computer, and I look at their results and decide when they are ready to go onto the next level.

Mrs. Rosen mentioned that she does not physically write her observations but makes mental notes of them. Ideally, Mrs. Rosen would like to use the following assessment techniques:

> Portfolio assessment, multiple measures, I would have some tests that I would give. Like math at the end of the chapter, I would have paper-pencil type tests, but I would also have the hands-on where they can show me: can they manipulate, can they explain? In writing, I would do samples, and look at their writing samples and see what elements they have. In reading I do like running records, and I do like comprehension questions that follow. I think that tells me enough through observation, too. I would do observational anecdotal records. A good teacher knows within a couple of weeks, what is

going on where that kid is. And that's what I would do: a variety of assessments.

Mrs. Rosen provided examples of how assessment techniques had helped guide her teaching. In response to my question about how standardized testing has impacted her teaching pedagogy, she declared that the assessment techniques develop a basis for focused teaching for her but that she is opposed to standardized testing. She especially emphasized how she has noticed a difference in the way she teaches math and writing due to the standards movement. In the lengthy excerpt below, the differences she has noted are presented in her own words:

> I think it has made a difference in how I teach, especially math. I don't think it has made a big difference in reading. At second grade, especially you are still teaching them foundational skills of how to read, and testing does not tell a whole lot about reading skills. Writing, I think that the way they assess the writing and six traits, has made a big difference in how I teach writing. It has given me a tool to teach writing actually. We are doing a lot more self-evaluation of our writing. The kids are looking at their writing with a different eye and thinking about their own evaluation, more meta-cognition about how they are writing, what they are writing.

She discussed how her focus is less on the genres and more on using the six traits to teach the genre. To explain her point she shared an example: "instead of teaching the genre of story, I teach the trait of voice and use story to show voice. Or when teaching the genre of letter writing, I teach the trait of idea and we use idea through letter writing." She continued to explain the impact on her math instruction:

> Math, I think that's where the biggest impact has been. I focus on one standard at a time. Like right now, we are doing number patterns and algebra. So, our activities are focusing on a lot of those things that is required by the district assessment plan. I know the statewide assessment instrument will eventually be tied to this. It [the standards] kind of guides what I am teaching. It doesn't really guide how I teach it. I really don't think much about Stanford-9. I don't think the test gives me accurate information about my students, it isn't intended to. If you really know what Stanford-9 is about, it's really intended to compare districts to districts. I am not threatened by my scores. My philosophy is if I have done a good job teaching, it will reflect that in the big picture. It is not a snapshot of whether I am a good teacher or not. I don't and won't be spending a lot of time on pre-testing. I might teach them how to bubble in their answers.

That's the skill that they need to know for the test. The test does not drive me. It is the learning that drives me and my instruction.

Author's Perspective

My teacher participants came from varied educational backgrounds. All of them had several years of experience working in the school district. They all had at least four years of experience teaching second grade. As mentioned earlier, the three schools differed greatly in the socio-economic population of students that they served.

There was a pattern in their responses about their teaching philosophies. One common theme was the fact that they believed that children learn at different paces and that instruction needs to be individualized and tailored to the needs of the students. One size does not fit all! Also, all my teacher participants pointed to the use of observation techniques to do their assessments. They also gave mini-tests at the end of units to evaluate the student's understanding of concepts and to know whether they are ready to move onto a new concept. These mini-tests helped them assess their own teaching and also student learning. Journal writing additionally played an important part as a tool for assessment in two of the three settings.

When asked to visualize an ideal classroom, I observed and recorded techniques that all three participants would like to use. They described how they would allow the students to use multiple methods of assessment, to check for understanding as opposed to a standardized method of bubbling in the right answer. All three teacher participants expressed their indifference toward the performance scores of their students on the Stanford–9 test and felt the scores were a mere snapshot of a single moment in their year of schooling, implying that the test does not take into account the student as a whole child.

The three teachers highlighted various developmentally appropriate practices in their teaching such as learning centers, small group instruction, cooperative groups, whole group instruction, and independent work, to name a few of the practices. All three of the teachers seemed to believe that there was no one right answer to a question and encouraged their students to demonstrate their learning using multiple modalities. In my observations, the teachers appeared to base their judgments about instruction on their students' learning. They also were eclectic in their style and could use a completely different teaching approach the following year based on different student needs in their classroom. Their experiences working in varied classroom settings and also their years of teaching experience enabled them to perceive their students' abilities and needs in making informed decisions about appropriate teaching techniques.

The fact remains that the three teachers taught in completely different settings. The standards they were required to teach were the same in all the three settings. The approaches and techniques that these three teachers used were varied and yet similar as this chapter makes apparent. For example, from my observations and interviews with the three teachers from different economic settings, I did not discern many differences in their attitudes toward the testing movement, based on the socio-economic status of their school settings. I did not notice much variation on the impact of standardized testing on teacher pedagogy based on socio-economic settings. What I did observe was differences in teaching styles, teacher-student interactions, which I would attribute to their individual teacher personalities more than socio-economic setting.

Concluding Comments

In the final analysis, the important thing about a test, which I learned in my research study in these second grade classrooms, is not its validity in general, but its validity when used for a specific purpose. It seems irrational that educational decisions that will have a major impact on students be based on a single test score, particularly when the test-takers are second graders like those I observed in the three different classrooms. Second graders are still developing foundational skills. For children who are limited English proficient, issues regarding language pose additional challenges. Other relevant information that teachers like Mrs. Mole, Mrs. Riser, and Mrs. Rosen gather as part of their pedagogical practices should also be taken into account. Lastly, for children below fourth grade—as was the case in my research—"the mechanics of taking tests and answering on specialized answer sheets can prove more difficult than the cognitive tasks the tests are asking them to address" (Haney cited in Peterson, 2000).

References

Atkinson, P., & Hammersley, M. (1994). Ethnography and participant observation. In N. K. Denzin & Y. S. Lincoln (Eds.), *Handbook of qualitative research* (pp. 248–261). Thousand Oaks, CA: Sage Publications.

Cronbach, L. J. (1990). *Essentials of psychological testing* (5th ed.). New York, NY: Harper & Row.

Erickson, F. (1986). Qualitative methods in research on teaching. In M. C. Wittrock (Ed.), *Handbook of research on teaching* (3rd ed., pp. 119–161). New York, NY: Macmillan.

Gardner, J. W. (1961). *Excellence: Can we be equal and excellent too?* New York, NY: Harper & Row.

Goodwin W. L., & Goodwin, L. D. (1993). Young children and measurement: Standardized and nonstandardized instruments in early childhood education. In B. Spodek (Ed.), *Handbook of research on the education of young children* (pp. 441–465). New York, NY: Macmillan Publishing Company.

Kohn, A. (2000). *The case against standardized testing: Raising the scores, ruining the schools.* Portsmouth, NH: Heinemann.

Meier, T. (2000). Why standardized tests are bad. In *Failing our kids: Why the testing craze won't fix our schools* (pp. 14–16). Milwaukee, WI: Rethinking Schools Ltd.

Meisels, S. J., Steele, D. M, & Quinn-Leering, K. (1993). Testing, tracking, and retaining young children: An analysis of research and social policy. In B. Spodek (Ed.), *Handbook of research on the education of young children* (pp. 282–289). New York, NY: Macmillan Publishing Company.

National Association for the Education of Young Children (1996). *NAEYC position statement on developmentally appropriate practice in early childhood programs serving children from birth through age 8.* Washington, DC: Author.

Peterson, B. (2000). Ban early childhood testing. In *Failing our kids: Why the testing craze won't fix our schools* (p. 41). Milwaukee, WI: Rethinking Schools Ltd.

Rudner, L. M., & Schafer, W. D. (2002). *What teachers need to know about assessment.* National Education Association of the United States.

Swope, K., & Miner B. (2000). *Failing our kids: Why the testing craze won't fix our schools.* Milwaukee, WI: Rethinking Schools Ltd.

Wortham, S. C. (2001). *Assessment in early childhood education* (3rd ed.). Upper Saddle River, NJ: Prentice-Hall.

Wortham, S. C. (2005). *Assessment in early childhood education* (4th ed.). Upper Saddle River, NJ: Prentice-Hall.

Summary and Implications

Susan McCormack
University of Houston–Clear Lake

Denise McDonald
University of Houston–Clear Lake

Tirupalavanam G. Ganesh
Arizona State University

Andrea S. Foster
Sam Houston State University–Huntsville

Testing and accountability continue to be an everyday concern worldwide. As evidenced by state mandates and actions of school districts, schools, and educators, testing and accountability have become significant aspects of contemporary American education. It is clear that test results are being used for multiple purposes and that these uses have extended well beyond the original purposes for which they were designed. Inferences made from test scores can be a means to improve education. However, it is important to remember that, "The only reasonable, direct inference you can make from a test score is the degree to which a student knows the content that the test samples. Any inference about why the student knows that content to that degree . . . is clearly a weaker inference . . ." (Mehrens, 1984, p. 10). Jacques Barzun in his foreword to the 1962 Banesh Hoffman tome, *The Tyranny of Testing*, reminds us that multiple-choice questions test nothing but passive-recognition knowledge, not active usable knowledge. Knowing something means the power to summon up facts and their significance in the right relations . . . mechanical testing does not foster this power (Hoffman, 1962). Nevertheless, indirect inferences from test results are made often and the evaluation of test results have been used to make decisions that affect every aspect of schooling from student achievement to teacher preparation and employability.

The five chapters in this section examined the impact and consequences of test results from a number of perspectives that include: evaluation of new teacher induction and mentoring programs; increased understanding of societal expectations for student outcomes; strong assimilation efforts toward Native populations; the "absent presence" in elementary school curricula; and the influence on teachers' pedagogy and the students' schooling experience.

Fletcher, Strong, and Villar use exploratory analysis techniques to establish a relationship between new teacher mentoring programs and the progress made in reading by students of new teachers. Whether or not student achievement data as measured by the Stanford 9 can be used for program evaluation purposes is contentious due to the complex nature of schooling and the need to account for a multitude of factors related to program implementation, mentor teacher characteristics, new teacher characteristics, curricular aspects, curriculum implementation facets, and student characteristics. Nevertheless, Fletcher, Strong, and Villar argue that yearly testing mandated by the NCLB legislation provides an opportunity that benefits educators. This study asserts that school districts need individuals on their staff who can provide research and evaluation support as related to the use of student achievement data and school accountability measures.

Lee's study of TIMSS data, although limited by the use of 8th-grade student achievement information and accompanying self-reported student surveys, provides an important examination of societal expectations for student outcomes in eastern (Japan and Korea) vs. western (England and U.S.) nations. This study indicates that high-stakes tests such as high school graduation and college entrance tests are double-edged swords—tools that when coupled with high academic standards have the potential to enhance student achievement and learning but also hurt students' self-concept and ability to perform well on the tests.

Klug's narrative suggests that NCLB regulation continues the governmental influence on Native American education. The push for high-stakes testing has a variety of effects on Native American students. Although reporting data to show student success keeps Native American students under public scrutiny, the discrepancies attributed to the ways yearly progress is reported result in underrepresentation of the Native American student populations. Further, the constant renorming process to align achievement with the dominant population contributes to significant scoring discrepancies and eventually leads to increased high school drop-out rates. Culturally relevant pedagogy is jeopardized. In many cases test anxiety on behalf of teachers and students facilitates the abandonment of culturally aware pedagogy. Furthermore, NCLB appears to be a contributor in the persistent struggle for control over education for Native Americans. Finally, stronger assimilation efforts accompany this legislation because of student performance requirements.

Snow-Gerono utilized survey narratives and focus group interviews to describe teacher mentor/interns' perceptions regarding changes in the classroom environment as a result of standardized testing requirements. More specifically, the research considers the impact of the high-stakes testing on preservice teacher preparation; on curriculum development and enactment in elementary classrooms; on the degree of the mentor teacher's job satisfaction; and, on how best to prepare new teachers for the standardization movement. This research considers the question "What is left out?" of standardized and test-driven curricula to be of great consequence. Teacher/educator voices are missing in most policy-making discussions that are aimed at choosing curricula and strategies. Of equal concern is the notion that standardization practices that limit differentiation opportunities create a "one-size-fits-all" mentality that largely ignores special needs and circumstances.

A. Ganesh investigates the impact of standardization on teachers' pedagogy and their students' schooling experiences through shared teachers' perspectives, experiences, and stories. The study uses interviews and observations to gain direct information on the plausible correlation between standardized test results and students' socio-economic status. A. Ganesh reports a myriad of teacher "voiced" observations/concerns/perspectives on how standardized testing does not support or reflect students' learning and growth, that it does not present useful information regarding student formative progress over a period of time, and that it serves as a skewed, overused, often misinterpreted "end all" for accountability. Collectively, the teacher participants in A. Ganesh's research believe that children learn at different paces, that instruction needs to be individualized, and that journal writing is a significant assessment tool.

While the debate about school accountability measures continues with the enactment of the NCLB Act of 2001, it is clear that there is a lack of consensus about desired educational outcomes. These chapters have provided insights into the complexity and controversy that surround both sides of the accountability challenge, particularly the high-stakes accountability debate. As a result, they have deepened our understanding of the value, the purpose, the intended positive, and the unintended negative effects of assessment systems in education today.

References

Hoffman, B. (1962). *The tyranny of testing*. New York: Crowell-Collier.
Mehrens, W. A. (1984). National tests and local curriculum: Match or mismatch? *Educational Measurement: Issues and Practice, 3*(3), 9–15.

Division 3
PERCEPTIONS AND PERSPECTIVES OF ACCOUNTABILITY SYSTEMS

Overview and Framework

Michele Kahn
University of Houston–Clear Lake

Mimi Miyoung Lee
University of Houston

Carrie Markello
University of Houston

Heidi C. Mullins
Eastern Washington University

Annapurna Ganesh
Arizona State University

> Michele Kahn is assistant professor of multicultural education at the University of Houston–Clear Lake. Her research interests include gender studies, teacher beliefs, and lesbian, gay, bisexual, and transgender issues.
>
> Mimi Miyoung Lee, Ph.D., is an assistant professor in curriculum and instruction at University of Houston. She received her Ph.D. in instructional systems technology from Indiana University in 2004.
>
> Carrie Markello, a Houston Endowment Fellow, works in the Laboratory for Innovative Technology in Education while pursuing a doctorate in art education at the University of Houston.
>
> Heidi C. Mullins is an assistant professor in art education at Eastern Washington University.
>
> Annapurna Ganesh, Ph.D., is a faculty member at the Mary Lou Fulton College of Education, Arizona State University.

Accountability in education, as readers learned in the previous sections, is a process by which certain parties are held responsible for reaching specific goals.

For example, teacher educators are deemed responsible for preparing "good" teachers, and teachers are expected to be responsible for preparing "good" students. The passing of the No Child Left Behind (NCLB) Act in 2001 dramatically changed the perspective of accountability in American education. Under this new lens, "good" means high scores on standardized tests. In this simple means-ends equation, teachers who are perceived as "good" have "good" students (high-scorers) and vice-versa. Underlying the NCLB policy is the assumption that if teachers are not performing (meaning that their students have low test scores), then sanctions need to be instituted to force compliance. If test scores or standards are not met, teachers and students are labeled as failing, which may result in public humiliation and/or diversion of funds.

Under NCLB, students, teachers, and teacher educators experience an assessment system that takes an often one-dimensional snapshot of their capabilities and potential. Key's (1998) use of the term, *pentimento*, to describe the changing perceptions of teacher interns is a useful metaphor for understanding how students, teachers, and teacher educators interpret and are influenced by assessment. A pentimento is "an underlying image in a painting, as an earlier painting, part of a painting, or original draft, that shows through, usually when the top layer of paint has become transparent with age" (The Free Dictionary, 2005). The current assumption underlying high-stakes testing is that an accurate and true evaluation can be made through a single quantitative instrument. The perspectives in the following chapters demonstrate that this is indeed a volatile and undependable measure from a human perspective. What then is the pentimento of our educational system if we were to examine what lays beneath high-stakes assessment measures?

The three authors featured in this section offer perspectives regarding policies and assumptions related to accountability issues in education. As in the pentimento metaphor, the issues surrounding accountability become transparent with time, inspection, and reflection, revealing underlying images that form the foundation for a contemporary picture of accountability. Kosnik's, T. Ganesh's, and Finnell-Gudwien's particular views on NCLB, standards, and high-stakes testing expose layers forming the current image of accountability. These authors offer multiple dimensions that both enlighten and call into question the assumptions and demands of accountability in education.

As a teacher educator and researcher who recently moved from Canada to the U.S., Kosnik describes her evolving views on research, NCLB, and the application of standards in her new preservice teacher education environment. Framed by a critical theory approach to meaning-making, T. Ganesh employs visual representations created and described by Arizona educators as an alternative window to their high-stakes testing experiences. Finnell-Gudwien focuses on "No Child Left Behind" through the lens of democracy in education as historically conceived. Presenting a theoretical framework based on Dewey's (1916) five tenets of democracy, Finnell-Gudwien systematically demonstrates gaps divorcing NCLB from the principles of democratic education.

In the current era of NCLB, educators from all backgrounds are jockeying to make sense of the implications of this mandate. Must one standard of testing apply to all? Are all students coming from the same knowledge base and cultural and ethnic backgrounds to merit this one-shot testing strategy? Of course the simple answer is no! Therefore, how do we address the need for standards in teacher education, teacher motivation, and student learning? These questions address extremely complex issues of quality or the "good" teacher and student, implementation of reform efforts, and the unique nature of educational settings. In spite of the complexity associated with these ongoing initiatives imposed on those involved in the education process, we must not only address the personal implications, but the social and longitudinal implications and the unintended ones as well.

Striving toward excellence often means using standards or benchmarks as markers of agreed levels of success, thus providing consistent achievement expectations. Standards, as Flinders informed us earlier in this volume, became a powerful movement at the turn of the century, creating a plethora of outcome-based assessments for teachers and students (Cochran-Smith, 2001). Overzealous use of standards in the name of school reform can result in de-humanizing the individual, as our first set of chapters suggested. If only measured by pencil and paper tests, the contexts and particulars of those who succeed and achieve in a multitude of exemplary ways are ignored and may appear to not meet the standards. Genuine standards recognize the importance of multiple perspectives and the complexities associated with teaching and learning.

The accountability movement's imposition of an overwhelming number of tests during the school year has taken charge of schooling. Assessments, such as portfolio reviews that measure progress of individual students over a period of time, can provide educators with valuable instructional information. On the contrary, tests that are taken once a year in a multiple choice format provide a mere snapshot of students' abilities. This type of assessment compares students with other students and offers little instruction with respect to where one should proceed from a learning perspective. The inherent danger of such assessments is that each student becomes a statistical number ignoring the environmental factors influencing the individual learner's learning. Furthermore, when school funds and status become tied to high-stakes tests scores, the educational environment and classroom climate risk being held hostage to short-sighted methods of accountability, as one of the visual images in this section suggests.

References

Cochran-Smith, M. (2001). Higher standards for prospective teachers. *Journal of Teacher Education, 52*(3), 179–181.

Dewey, J. (1916). *Democracy and education: An introduction to the philosophy of education.* New York: The Free Press.

Key, D. L. (1998). Interns' changing perceptions during internship (Doctoral dissertation, University of Alabama-Tuscaloosa, 1998). *Dissertation Abstracts International, 60,* 708.

The Free Dictionary (2005). *Pentimento.* Retrieved May 7, 2006, from www.thefreedictionary.com/pentimento.

CHAPTER 9

No Teacher Educator Left Behind

THE IMPACT OF CURRENT U.S. POLICIES AND TRENDS ON MY WORK AS A RESEARCHER AND TEACHER EDUCATOR

Clare Kosnik
Stanford University

> Clare Kosnik, Ph.D., is executive director of the Teachers for a New Era project at Stanford University. She is on leave from her position as associate professor in the Department of Curriculum, Teaching, and Learning at the Ontario Institute for Studies in Education, University of Toronto, and was previously director of the Elementary Preservice Program at OISE/UT. She chairs the Self-Study of Teacher Education Practices Special Interest Group of the American Education Research Association.

ABSTRACT

> This paper begins with a description of my work in a Canadian school of education, where I now realize that I was well insulated from the U.S. context, where many are highly critical of teacher education. My recent move to Stanford University immediately immersed me in the highly controversial practices being imposed on schools and schools of education through the No Child Left Behind (NCLB) legislation. I suggest that it was almost impossible for me to fully appreciate the impact of NCLB while working in Canada. I argue here that teacher education must maintain its integrity yet be mindful of the political agendas or risk being completely marginalized. I suggest that, while the development of self-study has been affected by the discourse and legislation, self-study can play a key

---

This chapter was originally published in *Studying teacher education*, vol. 1:2 (2005) and is reprinted with the permission of the author and Taylor & Francis, LTD., the website for which is www.tandf.co.uk/journals.

role in restoring balance to discussions of the importance of teacher education. In order to achieve this influence, we must continue to expand our research methods to enable us to participate in the language of evidence-based research.

Observing American Education from Canada

At the Ontario Institute for Studies in Education/University of Toronto (OISE/UT), I had the privilege of working in a fine public university that is able to offer high quality teacher education programs. OISE/UT has strong programs; however, it must continue to study its graduates and be mindful of new directions in teacher education. As I read research on teacher education and attend international conferences I am becoming increasingly aware of world-wide trends, perhaps the most alarming one being the effort to impose more control on teachers and in turn teacher education. Delandshere and Petrosky (2004) have noted such trends in teacher education:

> Movements to reform teacher education are underway in many parts of the world, including Europe, Australia New Zealand, and the U.S. These attempts at reform are motivated by various forces, but appear to reflect international convergence towards uniformity, conformity, and compliance. (p. 1)

For example, Australia has moved to generic teaching standards; New Zealand is raising standards and moving to a national curriculum; and Europe is standardizing teacher education through the development of common experiences and a course credit transfer system (Delandshere & Petrosky, 2004).

In Canada, education is a provincial matter with the federal government having very limited jurisdiction over it. However, there have been attempts in the last ten years by the Ontario government and its "arm's length" body, the Ontario College of Teachers, to determine program content and structure for teacher education and to impose an exit exam through the Ontario Teacher Qualifying Test (which was recently put on hold). Ontario has also begun to review teacher education programs for accreditation but not with the same degree of rigor as the National Council for Accreditation of Teacher Education (NCATE) process. Although teacher education programs and teacher assessment are much less prescriptive and regulated in Canada than in the U.S., the Canadian system in general seem to be moving somewhat in the same direction as other countries.

With the U.S. being our "neighbor to the south" and a leading force in research on teacher education, Canadians tend to watch closely its developing policies and practices. Having attended many conferences "south of the border," reviewed U.S. teacher education programs, conducted research on prominent American teacher education programs, and published mainly in U.S. journals, I naively believed that I recognized and appreciated the differences between the Canadian and U.S. systems. Being Canadian I felt somewhat insulated from U.S. Department of Education policies and initiatives. I saw them as seriously flawed (agenda of the right wing) and with some smugness could dismiss them. However, during the last few years I have noticed some changes in my thinking and practice. There was a turning point when I realized I had to rethink some of my research methods and reconsider my attitudes towards particular research, education policies, and practices.

In this paper I describe how I have grown to recognize that I did not really understand what I thought I understood. I chart some of the changes in my practice by identifying influences on my thinking. My recent move to Stanford University was only one catalyst for these changes. The lessons I learned might be useful for teacher educators both within and beyond the U.S.

Working as a Teacher in Canada

I see my research in Canada as falling into two major periods: the time before I completed a study on the admissions process to the OISE/UT teacher education program and the period after that study.

APPROACHES TO RESEARCH: PRIOR TO THE ADMISSIONS RESEARCH

In order to make its elementary program more "user friendly" and coherent, OISE/UT divides the 600 elementary preservice students into cohorts of approximately 60 students, each with a small faculty team and working with a limited number of practice teaching schools. As a beginning professor whose area of research was teacher education, I tended to focus on studying our own work and the progress of our student teachers. My early goals as an instructor in the program and later as a Coordinator of one cohort, Mid-Town, were to develop the program into an outstanding experience where students would pursue inquiry projects in their practicum placements; work in a cohort that was a strong, supportive community; and have a faculty team who developed a highly integrated program.

Using mainly qualitative research methods, but often supplementing interviews with surveys, I systematically studied and developed our program. With my Mid-Town colleagues we studied the practicum (role of the associate teacher, components of a good placement, and the use of faculty to supervise the practicum) and the academic program (use of action research, the curriculum, and foundations courses) and followed our graduates as they began their teaching careers.

In this period, there was a sense of urgency both to improve our teacher education program and to publish a significant number of articles in order to be granted tenure. In many ways, this period involved a constant stream of self-study projects aiming to understand teacher education through the eyes of our student teachers, the faculty in the program (tenure-line, contract instructors, and myself), and appreciate the challenges faced by beginning teachers. By systematically studying aspects of the program and its redesign I developed an in-depth understanding of Mid-Town. One of my biggest challenges was gaining approval from the university research ethics committee, which was initially reluctant to approve my studies because I was conducting research on my own students. Fortunately, this was sorted out and many others at OISE/UT followed my example.

When I became the director of the whole elementary teacher education program at OISE/UT, I extended my research to the entire elementary program, often involving other faculty and contract instructors in projects. I also branched out beyond my immediate university to look at the provincial level.

For example, working with a team of researchers we studied the impact of recent Ontario Ministry of Education curriculum reform on language arts and mathematics instructors at seven faculties of education in Ontario. Most reported there was limited impact, noting that they continued doing what they had done previously, namely, helping student teachers develop an approach to curriculum and teaching. In general they believed that students should be aware of the themes and specific outcomes in the formal curriculum but should learn to critique the curriculum and not be completely bound by it. Most instructors worked within the confines of their university and paid limited attention to Ministry of Education policies. These findings were consistent with my philosophy and practice and reinforced my beliefs about how to address the formal curriculum in teacher education.

In hindsight, I now realize that during this period I did not fully appreciate the trends in U.S. policies and politics. At the annual AERA conference I would listen attentively to my American colleagues lament the additional program requirements mandated by their state or the federal government, the challenge of applying for NCATE approval, the threat of alternative certification programs, and the emerging control of education by the right wing. Although I paid careful attention and offered words of encouragement, the true impact of these initiatives, policy requirements, and changing attitudes towards research eluded me.

THE ADMISSIONS STUDY: A TURNING POINT

I am very proud of the research I have described, believing that it led to improvements in our teacher education program and deepened my understanding of the process of becoming a teacher. I contributed to the growing body of literature on teacher education and used methods that matched the goals of the research. However, there was a turning point in my thinking and skills development that was spurred by a particular study I did on the preservice admissions process at OISE/UT. The next period led to a broadening of my perspective and a refining of my research skills.

OISE/UT is in both the fortunate and unfortunate position of receiving an abundance of applications for its teacher education programs. On average, 6,000 applicants seek admission either to the elementary or secondary program, hoping for one of the coveted 1,300 places. The majority are well qualified. The application process includes submission of a detailed statement describing three substantial experiences of working with students and explaining what was learned about the teaching/learning process through these experiences. This "profile" is heavily weighted. Transcripts must also be provided indicating their undergraduate GPA. It is worthwhile to note that unlike in the U.S., applicants to most Canadian schools of education are *not* required to submit standardized test scores (Praxis 1, SAT), and in all likelihood most applicants have never completed such tests.

Reading and assessing the thousands of applications is a labor-intensive and costly process. There have been repeated calls to either delete the written statement portion of the application or shorten it dramatically. As Director of the elementary preservice program and a committed teacher education researcher, I felt that any decisions on revamping the admissions process should be based on research. With the support and approval of the Registrar and the Preservice Admissions Committee, I set out with a colleague and a graduate student to do a study of our admissions process to determine its effectiveness, with a particular focus on the value of the written statement.

Although I knew that the admissions process was a controversial topic, I was shocked at the number of faculty who held extreme (and definite) views on how to conduct admissions. These ranged from interviewing all applicants to selecting students strictly on their undergraduate GPA. Almost immediately I knew this research would be closely scrutinized, perhaps in a way that I had never faced before. This led me and my team to focus heavily on study design and to include a strong quantitative aspect, which we believed was appropriate and would lessen allegations that our methods were subjective or not rigorous. We then expanded the research team to include OISE/UT research officers who vetted the design, completed some of the statistical analysis for us, and reviewed the text of some sections of the report.

Having used mainly qualitative research methods in the past, I found myself drawn to the quantitative part of the study. As we worked in SPSS doing crosstabulations I found myself wanting to "play" endlessly with the stats, which was not my typical response to this kind of data. The addition of the quantitative component strengthened our study and led me to a new appreciation of the information it provided. The two methodologies—qualitative (e.g. interviewing the cohort Coordinators about their students' performance in the program) and the quantitative (e.g. converting admissions information to numbers, ranking student performance in the academic program and practicum)—were complementary.

The study also introduced me to a body of literature that included extensive discussions of standards. As I began to pour over the literature on admissions, teacher effectiveness, and teacher characteristics, I began to take note of the emphasis on standards, which is not particularly prominent in Canada. I found this new literature fascinating and read extensively. The new body of literature and the research methods seemed to sit quite comfortably with my expanding philosophy.

The study on the admissions process was completed and the report presented to the Preservice Admission Committee. Although it has not so far had much impact on the admissions process at OISE/UT, an article based on the research has been accepted by a journal and hopefully will be of general interest to teacher educators. Further, it had a tremendous impact on me. I developed a new respect for quantitative research methods, was introduced to a new body of literature, improved my research skills, and gained confidence as a researcher. I needed these new skills and perspectives as I faced a new set of challenges.

Working in the U.S. Context

The study on admissions was followed by a dramatic change in my work. In January 2004 I was offered and accepted the position of Director of the Teachers for a New Era project at Stanford University. This put me squarely in the American context and immediately immersed me in the messy politics of teacher education in California, a state that is heavily regulated. As I stated earlier, I had not fully grasped the extent to which the U.S. teacher education "scene" was politicized and was "under siege." In this section I describe four key experiences: the first two required me to consider state and federal policy and the next two to focus on my growing awareness of shifts within my particular professional community. All four have contributed to a reassessment of some of my practices and beliefs.

NO CHILD LEFT BEHIND (NCLB)

If one were to ask a number of academics to identify pivotal events in recent U.S. education policy and practice, I would hazard a guess that the recent No Child Left Behind (2001) legislation would be noted as having a tremendous impact: "The enactment of the reform policies, their interpretation, and their consequences are visible at all levels of the educational system, defining the work of student teachers and teacher educators" (Delandshere & Petrosky, 2004, p.3). Although I had a fleeting knowledge of NCLB, I was wholly unprepared for its prominence in many (most!) discussions on teacher education and its influence in determining program development. Given that Canada does not have a federal department of education (or a national curriculum), initially I did not have a frame of reference to grasp the implications of NCLB:

> In an era of No Child Left Behind (NCLB) legislation, it is increasingly clear that teacher education programs will be held more accountable than ever before. Policies of the federal government demonstrate a fundamental mistrust of the field of teacher education, and the accompanying rhetoric implies that public schools fail, in large measure, because teacher preparation programs fall short in producing high quality educators (The Teaching Commission, 2004). (Post et al., 2004, p. 24)

As I am becoming quite familiar with NCLB, I am shocked at the ability of a government to impose practices (which are not universally accepted or based on research) on an entire nation and to force each state's "cooperation." For example, "in fall 2002, ECS created a 50-state data base (www.ecs.org/nclb) that tracks states' response to and compliance with the major provision of NCLB" (Coble & Azordegan, 2004, p.5).

U.S. teachers are attempting to work in an extremely rigid and punitive system that publishes schools statistics, student achievement, teacher bio data, and so on. Under the guise of student achievement—which "always seems to be the trigger for new educational reform policies"—in this reform effort "teachers are at center stage," being seen as "they are the main determinant of student achievement and learning" (Delandshere & Petrosky, 2004, p. 5). Using a very simplistic "logic," teachers are blamed for poor student performance and, in turn, their teacher education programs are deemed to be failures because they did not fully prepare teachers to be effective. Even in the face of substantial data on inadequate funding for schools and the impact of socio-economic inequities, the government has clung to this rudimentary analysis and its narrow, short-sighted dictums.

NCLB has caused me to quickly rethink my previous practice of dismissing most government policies. Working in the U.S., I cannot simply focus on developing a teacher education program based on principles of inquiry, integration, and community as I did with our Mid-Town program, even when I know these can and should be the framework for teacher education. The challenge is to prepare teachers to meet the state standards and still support their growth as thoughtful, reflective, and skilled practitioners committed to goals beyond raising test scores. If I do not want to be left behind I must be responsive to government expectations.

Accompanying the NCLB legislation was a strong move towards evidence-based research, usually conceived as quantitative in nature. The pages of *Educational Researcher* each month seemed to be filled with arguments for and against the valuing of only one type of research. The raging arguments gave me reason to pause and think seriously about methodology in a way I had not before. Since there is extensive discourse on this highly controversial topic, I only want to note that I am now acutely aware of the heightened emphasis on such issues.

REGULATING TEACHERS AND TEACHER EDUCATORS

At the American Educational Research Association's (AERA) annual meeting in 2004, I attended many sessions regarding teacher education in California. As I listened to reports and discussions I began to appreciate that I was leaving a setting that gave professors some level of independence and moving into unfamiliar territory, where educators were "experiencing intellectual de-skilling" (Sleeter, 2003, p. 23). Whereas standards are given a quick overview in most Ontario teacher education programs, they seemed to be the framework for many California programs; whereas I had quickly presented the formal Ontario curriculum (K to 8 in subject areas) to my student teachers, California student teachers were expected to pass standardized tests to prove their knowledge of content; whereas I had aimed for authentic assessment of my student teachers, the California credential system seemed remote and irrelevant to the real goals of teaching. I wondered how teacher educators could work in such a heavily regulated state.

For the admissions study I had read extensively about standards, but the actual implementation of them and their prominence in the day-to-day work of teachers and teacher educators had escaped me. I was beginning to realize that the development and implementation of standards for both curriculum and teacher performance were highly effective strategies for regulating teachers. I appreciated for the first time the significance of Robert Roth's (1996) descriptions of standards in teacher education and his observations and conclusions with respect to them:

> The domain encompassed by standards, accreditation, licensure, and certification is being reconstructed in fundamental ways. The impact is the creation of an entire historical era in the profession, equal in significance to other major periods in education history such as the development of normal schools. The standards movement is so pervasive and powerful that it may appropriately be termed the *Era of Standards*.
>
> The movement in general may be characterized by several salient features. Among these are a deep-seated and growing distrust of teacher education; a change in the locus of control, with national policy emerged as a dominant influence; restructuring of licensing and governance; and reconceputalizing the nature of standards, with performance and outcomes assuming a preeminent role. (Roth, 1996, p. 242)

It was not until I was immersed in and, to a degree, responsible for teacher education in the U.S. that I realized both the implicit and explicit control embedded in the standards. As I moved more fully into the U.S. and particularly the California context I was introduced to many new terms (credential, licensure) and a gaggle of acronyms (CSET, CBEST, TPE, CCTE, and RICA). As I learned what each one stood for, I saw a never-ending series of tests to regulate teachers. I now could appreciate Roth's observations that this was becoming an *Era of Standards*. I was truly shocked that California seemed to have standards for everything, with accompanying tests to determine whether the standards were being met. Given the insignificance of standardized tests in the Canadian system I am having to come to terms with the reality that I am working in a country that takes the completion of standardized tests as a given.

SHIFTS WITHIN THE SELF-STUDY OF TEACHER EDUCATION SIG

I have been a long-time member of the Self-Study of Teacher Education Practices (S-STEP) Special Interest Group (SIG) of AERA. Without question I see this SIG as my academic home and the members as my colleagues. Each year after AERA and on alternate years after the "Castle Conference" I felt renewed and returned to OISE/UT believing that we as a group were doing valuable work and making important contributions to the literature on teacher education.

Within this group I have begun to notice some shifts in our practice. At the AERA meeting in San Diego in 2004, S-STEP celebrated the publication of the *International Handbook of Self-Study of Teaching and Teacher Education Practices*. The two-volume work, addressing key issues in teacher education, seems to have moved us both symbolically and literally into a central place in the teacher education community.

The Handbook, which raises the profile of S-STEP, seems to coincide with a global trend to provide detailed documentation and rationales for particular practices in teacher education. The Handbook formalizes much of what has been learned about being teacher educators and details qualities of effective teacher education programs. This is a long way from our previous discussions where we grappled with questions about the self in self-study or "What compels us to focus on a personal understanding of our work as opposed to broader and loftier research issues?" (Freese, Kosnik, & Laboskey, 2000, p. 76).

At the *Fifth International Conference of Self-Study of Teacher Education Practices* at Herstmonceux, England in 2004, I noticed a few distinct changes to our processes. In the past, when proposals for this conference were reviewed, the reviewer often revealed him/herself to the author and the tone was very friendly and supportive. However, in 2004 all proposals were submitted to double-blind reviews. This system was put in place because some attendees could only receive funding from their universities if all proposals were treated in this arm's length manner. Granted that there was still the strong S-STEP commitment to collaboration and collegiality, I wonder if having an anonymous process changed the reviews.

In his opening address at the conference, John Loughran (2004) posed the following question about self-study: "How is the learning useable, applicable, and informing?" His question focuses on two central goals of the research, namely, to make a difference in teacher education and to demonstrate to others its effectiveness. He seemed to be moving self-study research beyond our local or individual contexts to make it accessible and usable to others.

In a similar vein, a plenary session led by Vicki Laboskey (2004) focused on research methods in self-study. She outlined five essential characteristics: self-initiated and self-focused; improvement aimed; employs multiple, mainly qualitative methods; interactive at one or more stages of the process; and validation achieved through the construction, testing, sharing, and re-testing of exemplars of teaching practice. Like Loughran, she seemed to be subtly moving self-study research into the more public forum where research processes are carefully scrutinized and "bump up" against those who hold narrow beliefs about research and adhere to quite conservative practices.

I wonder whether Loughran's and Laboskey's talks were somewhat influenced by the increasing emphasis on research methods. Although there were some shifts in the tenor of some discussions, I feel that we in S-STEP have stayed very close to our initial mission: to understand and improve our own practice within a community of critical friends. However, expanding the conversation as Loughran suggests may require us to consider how our work will be received by a larger audience.

MY MOVE TO STANFORD

While my appointment to Teachers for a New Era project at Stanford did not formally begin until August 2004, I began to be immersed in relevant information and issues several months before that date. Moving to another university and country I knew would involve a steep learning curve. I quickly realized that some of my philosophy and practices would have to change:

> The CSTPs [California Standards for the Teaching Profession] morphed from standards designed to promote teacher reflection to mechanisms for facilitating standardization, accountability, and summative evaluations, and what had originally been a holistic and multi-dimensional conception of teaching was reformulated as a linear and atomistic set of behaviors. (Berlak 2003, p. 33)

Similarly, during the AERA meeting, I seemed to be encountering standards at every turn. Sleeter notes that, in the new era, "the role of teacher education is to prepare teachers to teach the state-adopted content standards using state adopted materials, and . . . teachers will be evaluated based on their demonstration of competence in delivering this curriculum" (2003, p. 20).

My previous practice in Ontario of only briefly mentioning the standards would no longer be sufficient because they are central to the Stanford teacher education program. On a practical level I needed to learn the standards, and on a more theoretical level, I had to understand the implications of having them frame a program. Sutton, who is struggling with the problems of her minority students failing mandatory credentialing tests, highlights a particular concern of mine: "Adapting course content too closely to the test content also means that the test developers control the curriculum rather than educational institutions and faculty members" (2004, p. 468).

Not wanting to downplay how much I am having to learn about the California context, I feel this has been much more manageable and straight-forward than responding to the budding pressure to show the impact of teacher education programs on pupil learning. I have found the growing emphasis on value-added teacher education quite challenging:

> Value-added modeling (VAM) to estimate school and teacher effects is currently of considerable interest to researchers and policymakers. . . . They are particularly intrigued by VAM because of the view that its complex statistical techniques can provide estimates of the effects of teachers and schools that are not distorted by the powerful effects of such noneducational factors as family background. (McCaffrey et al., 2003, p. iii)

Since Stanford University is a Teachers for a New Era (TNE) site, we are being required to show definitively, using quantitative research methods, the link between teacher education programs and pupil progress. I am not sure what is more daunting, having to control for all the variables in the study or being required to use only quantitative measures. On a daily basis I struggle with this aspect of my new position.

Stanford is a research-intensive university par excellence. Working alongside Linda Darling-Hammond, a quantitative researcher who specializes in research design and has numerous large-scale projects that use massive amounts of survey data, has led to an increased awareness of quantitative study design. I am glad that I had the experience with the admissions study, which required us to think very seriously about quantitative design and led to us working with two OISE/UT institutional researchers who specialize in quantitative methods. If I am to succeed as Director of Teachers for a New Era I need to expand my repertoire of research methods, particularly as applied to large-scale studies. I have been feverishly reading about quantitative research methodology and studying examples of large-scale projects.

Balancing My Integrity and Not Being Left Behind

The previous discussion identified some emerging trends. It has become strikingly apparent that I must continue to modify my practices, or I will be left behind. The challenge will be to remain true to my values while living and working in a highly politicized world. This will require a high level of ingenuity and the identification of strategies to achieve this balancing act.

DEVELOPING A KNOWLEDGE BASE FOR TEACHING AND TEACHER EDUCATION

Cochran-Smith in her editorials for the *Journal of Teacher Education* often provides a balanced perspective on the overheated discourse on teacher education. She can cut through the hyperbole and emotion to focus on the key issues. In her March/April article I felt that she identified a strategy we might want to consider seriously: "in many of the major 21st-century debates about teacher quality and teacher preparation, the central focus, at least on the surface, is on research itself, particularly on whether there is a research base for teacher education" (2004, p. 111). Her emphasis on research resonated with much of what I have been experiencing—the need for strong data to support our program choices and practices.

Darling-Hammond, who has written extensively on educational policy, comes at the issue of a research base from a different angle but reaches a similar conclusion:

> Policies that support teachers' professional learning can make a major difference in student success. But figuring out what kinds of policies support teaching that meets today's new and very different goals for students is not at all simple. (1997, p. 35)

In another text, writing with Wise and Klein, she describes the situation as follows:

> Over the years, the occupation of teaching has had difficulty defining standards and enforcing a common knowledge base. This has been the case partly because, in contrast to other professions that have developed throughout the twentieth century, teaching has been governed through lay political channels and government bureaucracies—state legislatures along with state boards and departments of education—rather than professional bodies charged with articulating and enforcing knowledge-based standards. (Darling-Hammond, Wise, & Klein, 1995, p. 7)

Through various publications—handbooks, texts, articles—we are compiling a substantial set of research studies on teacher education. These will be crucial in helping us develop a knowledge base for teachers and teacher educators, one that has applicability in multiple contexts and can be used with different audiences (i.e. government agencies, policy think-tanks). *The Handbook of Research on Teacher Education* is a fine volume, but we need to go beyond it to one that includes more specifics regarding practice. I fully recognize this is not a simple task, but I see the need for us to develop our practice rather than have an inappropriate one imposed on us.

I found a recent conversation with a dean of a very large teacher education program quite alarming, but it illustrates clearly why we need to use our research to define what is a good teacher education program, what is effective practice, and how to help beginning teachers become highly skilled. He outlined some strategies he was implementing to "improve" the preservice program at his university:

- Group preservice students (at least 300) in the lecture hall for lectures delivered by tenured professors (many of whom have no recent experience in schools)
- Do away with practicum supervision
- Focus strictly on theory

His suggestions were extreme and should have been immediately discarded. Just as no reputable doctor would prescribe sugar as a mainstay for a diabetic, no Dean of Education should impose such wrong-headed practices on a program. We need to be able to immediately counter such inappropriate suggestions with data to "prove" these methods would be harmful to teacher development, and that student teachers would not acquire in this manner the knowledge, skills, and attitudes required of them.

Compiling a knowledge base for teachers is daunting, but finding ways for students to learn it, often in a one-year program, is almost overwhelming. In a discussion with Lee Shulman about the challenges of teacher education, he offered a suggestion that I found highly interesting. He proposed that we look to the preparation of lawyers as an example. He noted that almost every law school uses one of three or four sets of cases that are respected and valued by the profession. He described the need for a common set of case studies to be used in our teacher education program. I am not sure what this might mean for teacher education and how it would be done but it merits close consideration.

Much to my surprise, I am beginning to understand the place of standards for teachers and for curriculum. Standards for teachers, if appropriately formulated can provide a rationale for teacher education and describe appropriate practice. In order to acquire the knowledge, skills, and attitudes of thoughtful and effective practitioners, beginning teachers need specific knowledge; they require time to learn their craft; and they need to be enrolled in substantial (not shortened) programs developed by skilled teacher educators.

My discomfort with standards has not vanished entirely, but I can see their value if they are (developmentally) appropriate and if they are used in a way that helps student teachers become effective teachers. If they are simply lengthy lists of inappropriate or vague objectives, they can easily smother the inquiry and reflection elements of teacher education. But if they are thoughtfully developed, used judiciously, and not tied to standardized tests, they could bring coherence to a program. This in turn could also bring some consistency to teacher education. Programs that use standards based on research may provide us with the arguments to refute many of the common criticisms leveled at teacher educators.

CROSS-SITE RESEARCH

In order to continue to develop our knowledge base, I would like to see different schools of education collaborating on research, studying the same questions, using common instruments, and contributing data to a common database. The benefits of this type of collaboration are many: our collective voice would be strong; our policies could be based on large-scale research; and our colleagues in smaller or teaching-focused universities where research is more difficult to conduct could still contribute to the research and access the data.

In the Teachers for a New Era program in which I am involved, the 11 member schools are developing programs that "should be guided by a respect for evidence. A culture of research, inquiry, and data analysis should permeate the program" (Carnegie Corporation of New York, 2001). This select group must expand dramatically to engage with many others.

An example we might want to consider is the Performance Assessment for California Teachers (PACT) Consortium in California. I am seeing the tremendous benefits of the 16 schools of education (and increasing each year) using the same capstone experience (i.e., a portfolio—teaching event and embedded signature assessments, rather than an ETS developed standardized test). In addition to the common assignment, there is on-going research, during the credential year and afterwards, to study the effects of the teacher education program on teachers' practices. This joint endeavor is a clear demonstration of the power of collaboration. Few of these universities could have developed PACT and implemented its accompanying research on their own, but a consortium has the person-power and funds to implement innovative projects of this kind.

I believe that, unfortunately, the value-added trend is not a fad that will quickly be relegated to the dustbin. Further, there is a kernel of truth within the VAM which we should consider. We need to show that our quality teacher education programs make a difference in the lives of children. Student progress should not be determined by a single score on a standardized test; rather, we need to document student progress in a way that is holistic and yet not totally subjective. Academics, especially those in research-intensive universities, need to show the benefits of investing in teacher education; they need to establish that in the long-term, student learning will be greater if they are taught by well-prepared teachers. We must show the problems with the assessments advocated through VAM, but at the same time offer alternatives that are supported by research. Yes, this is a tall order, but working together I believe we can show that a more holistic and teacher-developed assessment process has much merit and validity.

An Even Stronger Role for S-STEP

As I work in the overheated California context I am more fully convinced that my S-STEP colleagues can and should have a greater role in the broader discussions about teacher education. Ken Zeichner's (1999) endorsement of S-STEP is encouraging and heartening in these sometimes dark days for teacher education:

> Researchers in the self-study movement in teacher education have employed a wide variety of qualitative methodologies and have focused on many different kinds of substantive issues. . . . A whole group of self-studies focuses on the tensions and contradictions involved in being a teacher educator in institutions that do not value this work. . . . Much of this work has provided a deep and critical look at practices and structures in teacher education. (Zeichner, 1999, p. 11)

S-STEP members tend to be key individuals within their universities who are involved in the daily (and often labor-intensive) work of teacher education. They understand the issues and have valuable suggestions on what needs to be done. The flurry of publications from S-STEP SIG members must not subside. For example, Garry Hoban's (2005) text, *The Missing Links in Teacher Education: Innovative Approaches to Designing Teacher Education Programs*, provides many examples of program development based on research. The new journal, *Studying Teacher Education: A Journal of Self-Study of Teacher Education Practices*, is peer-reviewed and promises to disseminate the work of self-study to a broad audience.

Conclusion

Over the past five years, some of my perspectives and practices have changed. I now give a greater attention to the global context and am acutely aware of how difficult it can be to gain a true understanding of another culture when working outside of it. Shared research agendas, collaboration across countries, joint publications, exchange programs, and on-going conversations may be strategies to help acquire insider knowledge.

As a researcher I know that I am stronger as a result of my recent experiences. I have learned a great deal and continue to hone my skills. I value quantitative research methods more fully and recognize how qualitative and quantitative methods can work together. My greater attention to research design has improved the quality of my work, which in turn benefits the teacher education programs with which I am involved because I have better data and more thorough analysis.

With research in hand, we must continue to share our work with a larger audience, even those who are quite hostile to it. If we can learn to speak "some of their language," which is often in quantitative terms, we will have a greater chance of having an impact.

Not all the trends I described earlier are inherently problematic. Some have such negative overtones that it is easy to dismiss them completely, but I think we need to look for what is of value in them. More than ever we need to be aware of the politics of education and be prepared to meet in the political arena, even if it is often skewed against those of us researching and practicing teacher education.

As director of the elementary teacher education program at OISE/UT, where we had 11 cohorts, I encouraged each team of instructors to individualize their program. I would never want to discourage faculty from bringing forward their special talents and interests. However, I think we need to balance individuality with consistency. We are in a highly politicized field that is being scrutinized. I am moving slightly in the direction of having more consistency in our programs.

References

Berlak, A. (2003). Who's in charge here? Teacher education and 2042. *Teacher Education Quarterly, 30*(1), 31–40.

Carnegie Corporation of New York, July 1, 2001 Announcement, A National Initiative, Teachers for a New Era from www.Carnegie.org.

Coble, C., & Azordegan, J. (2004). The challenges and opportunities of the No Child Left Behind act: Seven strategies for teacher educators. *Action in Teacher Education, 26*(2), 2–14.

Cochran-Smith, M. (2004). Ask a different question, get a different answer. *Journal of Teacher Education, 55*(2), 111–115.

Darling-Hammond, L., Wise A., & Klein, S. (1995). *A license to teach: Building a profession for 21st-century schools.* Boulder: Westview Press.

Darling-Hammond, L. (1997). *The right to learn: A blueprint for creating schools that work.* San Francisco: Jossey-Bass.

Delandshere, G., & Petrosky, A. (2004). Political rationales and ideological stances of the standards-based reform of teacher education in the U.S. *Teaching and Teacher Education, 20,* 1–15.

Freese, A., Kosnik, C., & LaBoskey, V. (2000). Three teacher educators explore their understandings and practices of self-study through narrative. In J. Loughran & T. Russell (Eds.), *Proceedings of the Third International Conference on Self-Study of Teacher Education Practices* (pp. 75–79). August 29–31. East Sussex, England.

Hoban, G. (Ed.) (2005). *The missing links in teacher education: Developing a multi-linked conceptual framework.* New York: Springer.

Laboskey, V. K. (2004, June/July). Talk on new directions for self study. Presented at the Fifth International Conference of Self-Study of Teacher Education Practices, Herstmonceux, England.

Loughran, J. (2004, June/July). Informing practice: Developing knowledge of teaching about teaching. Symposium presented at the Fifth International Conference of Self-Study of Teacher Education Practices, Herstmonceux, England.

McCaffrey, D., Lockwood, J., Koretz, D., & Hamilton, L. (2003). *Evaluating value-added models for teacher accountability.* Santa Monica, CA: Rand.

Post, D., Wise, K., Henk, W., McIntryre, J. D., & Hillkirkm, R. K. (2004). A new pathway for the preparation of highly qualified teachers: The master of arts in teaching (MAT). *Action in Teacher Education, 26*(2), 24–32.

Roth, R. A. (1996). Standards for certification, licensure, and accreditation. In J. Sikula, T. Buttery, & E. Guyton (Eds.), *Handbook of research on teacher education,* 2nd Edition (pp. 242–306). New York: Macmillan.

Sleeter, C. (2003). Reform and control: An analysis of SB 2042. *Teacher Education Quarterly, 30*(1), 19–30.

Sutton, R. (2004). Teaching under high-stakes testing: Dilemmas and decisions of a teacher educator. *Journal of Teacher Education, 55*(5), 463–475.

Zeichner, K. (1999). The new scholarship in teacher education. *Educational Researcher, 28*(2), 4–15.

CHAPTER 10

Critique through Visual Data
THE IMPACT OF HIGH-STAKES TESTING
EXPRESSED THROUGH TEACHERS' DRAWINGS

Tirupalavanam G. Ganesh
Arizona State University

> Tirupalavanam G. Ganesh, Ph.D., is a principal research analyst for the Office of the Dean at the Mary Lou Fulton College of Education at Arizona State University, Tempe. He is a December 2003 graduate of the Interdisciplinary Ph.D. program in educational media and computers, division of curriculum and instruction, at the College of Education, Arizona State University. His research interests are in visualizing data, communication of research, and science, technology, engineering, and mathematics in education.

ABSTRACT

This chapter offers a content analysis of selected visual representations of educators' feelings about the implementation of a state-mandated high-stakes test. Infused with perspectives from critical theory, I have employed visual data to elicit educator's views of their condition in an increasingly mechanized enterprise where teachers are expected to do what they are told with no room for using their intelligence, creativity, and judgment. The management of educational practices using the technology of high-stakes testing suggests an attempt to control a largely complex, dynamic, evolving, and human enterprise with rules, formulae, and algorithms.

In the United States, with the introduction of state-mandated school accountability measures and the implementation of the No Child Left Behind (NCLB) act of 2001, high-stakes tests are widely used as promotion and graduation requirements (Horn & Kincheloe, 2001). "High-stakes testing" has been defined by Swope & Miner (2000) as follows:

> When an educational decision is based on a single test score—whether a student will advance to the next grade level, be able to enter a preferred program or school, or even get a high school diploma. High-stakes are also applied to schools and teachers, with judgment, rewards, or punishments, based wholly or primarily on standardized test scores. (p. 140)

Tests enable the local, national, and international comparisons of curricula, students, teachers, and schools; the prescriptive reforms to "fix" American schooling; and the political movements to denigrate and destabilize public schools so as to glorify the import of the free-market notion to education. The very notion of standards and standardized tests confirm that schools do not exist in isolation from larger society. Giroux (1988) emphasized that "[s]chools must be seen as institutions marked by the same complex of contradictory cultures that characterize the dominant society" (p. 7).

Theories of social reproduction in education, though differing in important aspects, fundamentally propose that educational institutions play a significant role in social and cultural reproduction (Morrow & Torres, 1995). Schools are also some of the most important ideological machines of the state. What are the state's interests in education—democratic equality, social efficiency, and social mobility? According to Labaree (2000), these goals for education conflict with one another and priorities shift over time. What schools do ideologically, culturally, and economically is extraordinarily complex and may not be easily understood by the application of a simple formula.

There are very strong connections between the formal and informal knowledge within the school and the larger society with all its inequalities. Daily happenings in school are intricately connected to patterns of disparity found in the economic, cultural, and political aspects of society (Apple, 1990). Schools are social sites comprised of complex, dominant, and subordinate cultures, each characterized by the power they have to define and legitimize a specific view of reality (Giroux, 1988). Intertwined in the complex reality of school life, which is an integral part of our social fabric, are teachers' views and feelings about high-stakes tests.

This chapter presents a select few visual representations of Arizona educators' feelings about the Arizona Instrument to Measure Standards (AIMS) test and an analysis of these images. The AIMS is a competency test that students must pass to graduate from high school. It is based on the Arizona Academic Standards, which are skills in reading, writing, and mathematics that students statewide are expected to master.

In 1996, the Arizona legislature passed a law directing the State Board of Education to "develop and adopt competency tests for the graduation of people from high school." This law also stipulated that the State Board establish passing scores for each test. Originally intended as a graduation requirement for the class of 2000, this test as a graduation requirement was postponed more than once before the new deadline of 2006 was set (see www.ade.state.az.us/AIMS/).

The spring 2000 administration of AIMS resulted in eighty-three percent of the sophomores taking the exam failing the mathematics portion of the test. Results of the spring 2000 administration of AIMS were printed in *The Arizona Republic* (Flannery & Pearce, 2000), thus enhancing the high-stakes nature of the test. AIMS's results were also reported by state, county, and school district, indicating the percent of students who "Fall Far Below the Standard, Approach the Standard, Meet the Standard, and Exceed the Standard." The September 28, 2000, edition of *The Arizona Republic* reported pass rates for Phoenix Valley sophomores in the mathematics, reading, and writing portions of the AIMS under the heading "Best, worst performers." The socio-political nature of the AIMS test was obvious; the then Superintendent of Public Instruction, Lisa Graham Keegan's position was explained as follows:

> Keegan said parents of students who failed must recognize that "time has run out on hoping this will be taken care of" and must demand help from their schools. . . . She strongly defended keeping AIMS as a graduation requirement. Without a penalty for poor scores, she argued, students and schools have no incentive to take tougher standards seriously. (Flannery & Pearce, 2000)

From 2001 to 2003, I attempted to gain access to urban and suburban high schools in Phoenix, Arizona, where I hoped to use qualitative research methods (Erickson, 1986) to understand the high-stakes nature of AIMS and its impact on the curriculum of school. Public school administrators refused to take part in a long-term qualitative study, expressing discomfort with the topic due to the socio-political nature of the test. Some educators expressed their reluctance to speak verbally about the AIMS in their school settings. Teachers felt that they could be reprimanded by school administrators and expressed concern that they could "get in trouble."

Yet, in such an atmosphere pervaded by fear, there were many educators willing to make a drawing that expressed how they felt about this high-stakes test. I collected over 80 drawings created by teachers from around the state. These teachers also shared verbally in an interview and in short written descriptions of their drawings what they attempted to convey through their visual representations.

The primary concern that leads to disquiet with an endeavor such as this, where visual data were collected, rests with the nature of the response, visual versus verbal. The relative familiarity of researchers in the field of education with verbal and textual data places the researcher using visual data in an untried arena. Nevertheless, the drawings represented a corpus of data that allowed for sense making in ways I had not encountered previously—I was faced with meaning-making and interpreting educators' drawings.

The visual representations themselves had their own effect and therefore needed careful examination. The *ways of seeing* images are significant in understanding the socio-political, ideological, and cultural aspects of the production and re-presentation of visual representations of the teachers' high-stakes testing experience. The analysis of visual data using interpretive techniques is challenging—interpretations are limited to the context. However, the analytic process offers the potential for multiple plausible explanations. The possibility of multiple explanations based on the diverse perspectives of various participants, the maker of the visual images, the researcher, and the readers make the analysis of visual data complex and inclusive. The unique perspectives and world-views that readers bring to bear when reading images overlap with the context of the production and reproduction of images.

In consideration of these notions, I focused on seeking out the meaning of the visual representations under study and the articulation of meanings about the lived and perceived worlds of participants. Here, I use the term participants in its expansive sense to include the creator, the researcher, and all readers of the images. Drawing from readings on critical theory, I developed a focus for making sense of the visuals in the context of research in education.

Applying the lens of symbolic analysis to visual representations (i.e., the drawings) made by Arizona educators regarding their feelings about high-stakes testing, I found that their representations reveal exceptional interpretations. Meanings of symbols embedded in educators' drawings make the familiar strange and complex again. The representations themselves evoke visceral responses in the viewer, bringing to bear the viewer's own socio-cultural and world-views of the phenomenon of education.

Educational research in which visual data are used critically, or presented as data and not merely as visual aids, is infrequent to say the least. When students in Massachusetts were asked to draw and describe their reactions to their state's competency tests, students expressed concern about their performance and described feelings of frustration (Wheelock, Bebell, & Haney, 2000a, 2000b).

Popkewitz (1999) advocated for researchers to critically inquire and reflect on what one sees, to examine how those images are constructed and reconstructed. When considering the value of visual images as data in educational research and when attempting to access their representational value, it is necessary to consider the cultural, social, and economic conditions surrounding the creator of the images, the educational researchers, and readers (Fischman, 2000).

Attempts at critical incorporation of visual data and methods in educational research are not intended to create a "true" representation of phenomena; rather, they are aimed at "viewing" education from another lens without throwing away the established approaches in the field. The reliance of educational research on sources of information that can be quantified and on the spoken and written word cannot be abandoned and replaced by the sole use of visual data and related methodologies. Images, photographs, and drawings are constantly in need of interpretation and sense-making that is resistant to mere data reduction techniques. In this research endeavor, I explore the use of visual data as a distinct tool in educational research.

Meaning-making is not merely a process of categorizing and analyzing content for certain pre-specified aspects—which in itself is challenging. It is also a process that includes developing a critical understanding of the visual representations in the context of their creators' lived lives. More specifically, meaning-making involves exploring one's reactions to the visual data in relation to the structural conditions in which schools and curricula function. I invite the readers of this chapter to explore the following images with a critical eye.

Powerless

The central problem in "reading" the self-portrait of a female teacher in figure 10.1 is the contrary notions of powerlessness, isolation, and estrangement—the teacher is a prisoner, yet rebellious, creative, and iconoclastic (Ganesh, 2002). The basic feeling evoked is that of alienation—that of divestment. Reitz (2000) described alienation as inseparable from our enduring inquiry into social life (p. 51–77).

The author of this powerful image said, "I have immense feelings of frustration watching children struggle on a test that will define their future." In an interview, she expressed her feelings of anguish that her power as a teacher was forcibly removed by others outside of her classroom. She said about her visual representation of her feelings about the AIMS test, "The teacher [in this image] has blinders because she does not know what's on the test and she does not know how the standards are measured. She is gagged and shackled because she can't help the student bring out the learning she knows that student has, but is not accessing on a fill in the blank test."

This visual image, representative of a teacher held hostage by a high-stakes test, signifies this teacher's symbolic resistance to a state-mandated test. Her act of making such a drawing is a silent protest at the "deskilling" of her profession

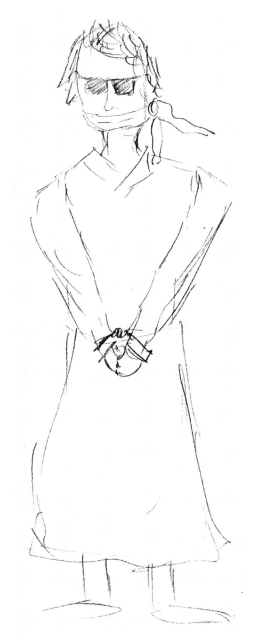

Figure 10.1 Powerless

by individuals far removed from the life of her classroom. This teacher said she felt helpless and powerless. Yet, the depiction of a teacher as a hostage is an act of iconoclasm—the mythic image of the revered teacher is thus shattered. This portrayal invokes the image of a female educator's loss of control as a facilitator of learning and embodies her sentiment as a hostage and a victim.

The Modern Sisyphus

The male teacher's representation in figure 10.2 is evocative of the Greek mythical figure Sisyphus trying ceaselessly to roll a rock uphill which forever rolled back on him. The inclined ground—overlaying the male English educator's familiar world of literacy—makes his very foundation slippery. According to this teacher, the state-mandate of the high-stakes test as a graduation requirement was forever crushing him in his role as a teacher. He expressed that teaching under such conditions was a dreadful form of punishment that made teacher labor ineffective and hopeless.

The portrayal of the teacher as the masculine punished hero is particularly tragic because this image demonstrates that this teacher is conscious of his situation and the conditions under which he teaches. Where would this teacher's distress at his situation be if he were confident that all of his students would pass the AIMS test? He labored every school day at the same tasks, consciously knowing the absurdity of the situation—the very premise of the AIMS test incorporated the creation of a class of winners and losers.

This English language arts teacher said, "The very foundation I attempt to build each day with my students is threatened by high-stakes testing. I work to facilitate a learning environment where my students learn to think critically and question the status quo." However, he expressed that this idealism about his role was marginalized due to the pressure from his community and the media. He said, "I start out each year with these seemingly lofty ideas about educating individuals for a democracy. But, there is this tremendous and constant pressure from the school administration and the media to magically increase our AIMS pass rate." Many teachers in the study concurred that their teaching had become increasingly dominated by someone else's conception of what must be done in the classroom.

No Creative Teaching

The feeling that high-stakes tests devalue teacher autonomy and that the pressure to raise test scores puts forward rote learning, valuing drill and kill work-

Figure 10.2   Modern Sisyphus

sheet activity over spontaneity and creativity, comes through in the female teacher's depiction shown in figure 10.3. Characterizations of teacher labor as cheap—evidenced by low pay, as less respectable, and open to more social blame, and evidenced by the ease with which various socio-political groups ascribe low test scores simply to "poor" teaching—take new meaning when we see this representation. Clearly, this teacher felt that the Arizona state standards and the AIMS test as interpreted by the state department of education representative who spoke to educators at a district-wide meeting were very prescriptive. She said that her drawing illustrates the plight of many teachers:

> That is where the AIMS test makes me so uncomfortable. Suddenly somebody—*they* are saying that many of your students won't graduate because you must have done a terrible job of teaching them. But that just isn't so. A good teacher relates to her students and does the best she can.

She also expressed concern about the prescriptive nature of the many messages she was receiving:

> The state department of education representative and my school administrators tell us teachers that we have to do these specific things in our classrooms to get the test scores up. I have even been told that it would be better if my students sit in individual desks rather than work collaboratively. We are not allowed to go on field trips unless the purpose of the field trip directly relate to areas that will be tested in the AIMS.

This 22-year veteran language arts teacher was protesting the fact that her school administrators were of the view that somehow changing the classroom seating plan would be helpful in bettering student achievement as measured by the AIMS test. She also felt pressured to switch to a traditional lecture format of teaching from her uniquely creative style which focused on individual students' needs. Implementation of the state-mandated policy on school accountability, according to many teacher participants of this study, resulted in ineffective teaching practices, thus reinforcing Swope and Miner's (2000) argument that the "misuse of standardized tests can distort learning, exacerbate inequities, and undermine true accountability" (p. 8).

It is important to note that the notion of school accountability most frequently evoked—by the media, politicians, and the general public—seems to be units of measure that make it possible to quantify the performance of students, teachers, and schools. However, teachers and students comprise concrete individuals who labor in an institution called school. The overriding culture of accountability views these concrete individuals through numerical measures im-

Figure 10.3 "No creative teaching"

plying a belief in objectivity, the ability to quantify the performance of students, teachers, and schools (Eisner, 1998). Our nation's faith in standardized tests underlies our confidence in "right answers" and belies the value of a culture that engenders genuine intellectual exploration—what Greene (1995) called the releasing of one's imagination.

Teachers in this study articulated that the high-stakes testing movement had affected them psychologically and emotionally. One teacher said, "Education is more than rote learning; education involves a whole person. For that, you need a teacher, a human person who cares about her students and the people who care, need caring too." The symbolic meanings in educators' drawings presented here reveal how pride in the teacher's own work is endangered by the high-stakes movement. Conversations with teachers in this study confirmed that they were conscious of the rapid pace of the deskilling of their own profession. They also expressed feelings of fear, anger, hopelessness, powerlessness, and frustration.

The wealth of discussion the production and re-presentation of educators' visual representations regarding the Arizona Instrument to Measure Standards (AIMS) test or any high-stakes test could engender includes the potential to reveal the hidden as well as commonly held meanings about schools, teaching, curricula, policy, and the role of formal education. Visual methods offer an alternative way to access educator's thinking and feelings about a high-stakes test like the AIMS.

The visual representations are a tool to educate, ask questions, and reflect upon the condition of education. The engaging nature of using visual representations in the sense-making processes of understanding schooling and education in all its aspects makes the consideration of visual methods interesting. Teachers' abilities to visualize what they see in education—in this case, their experiences of high-stakes accountability testing—and the ways they see it should not be overlooked as powerful data sources of the impact of government-mandated testing on those closest to students—i.e., those most likely to influence students' test scores.

References

Apple, M. W. (1990). *Ideology and curriculum* (2nd ed.). New York, NY: Routledge.
Eisner, E. W. (1998). *The kind of schools we need: Personal essays.* Portsmouth, NH: Heinemann.
Erickson, F. (1986). Qualitative methods in research on teaching. In M. C. Wittrock (Ed.), *Handbook of research on teaching* (3rd ed.) (pp. 119–161). New York, NY: Macmillan.

Fischman, G. E. (2000). *Imagining teachers: Rethinking gender dynamics in teacher education.* Lanham, MD: Rowman and Littlefield.
Flannery, P., & Pearce, K. (2000, September 28). Sophomores' scores improve but most still fail. *The Arizona Republic,* p. A1, A12–13.
Ganesh, T. G. (2002). Held hostage by high-stakes testing: Drawing as symbolic resistance. *Teacher Education Quarterly, 29*(4), 69–72.
Giroux, H. A. (1988). *Teachers as intellectuals: Toward a critical pedagogy of learning.* Granby, MA: Bergin & Garvey Publishers, Inc.
Greene, M. (1995). *Releasing the imagination: Essays on education, the arts, and social change.* New York, NY: Jossey-Bass.
Horn, R. A., & Kincheloe, J. L. (2001). *American standards: Quality education in complex worlds.* New York: Peter Lang.
Labaree, D. F. (1997). Public goods, private goods: The American struggle over educational goals. *American Education Research Journal, 34*(1), 39–81.
Morrow, R. A., & Torres, C. A. (1995). *Social theory and education: A critique of theories of social and cultural reproduction.* Albany: State University of New York Press.
Popkewitz, T. S. (1999). A social epistemology of educational research. In T. S. Popkewitz & L. Fendler (Eds.), *Changing terrains of knowledge and politics* (pp. 17–42). New York, NY: Routledge.
Reitz, C. (2000). *Art, alienation, and the humanities: A critical engagement with Herbert Marcuse.* Albany: State University of New York Press.
Swope, K., & Miner, B. (2000). *Failing our kids: Why the testing craze won't fix our schools.* Milwaukee, WI: Rethinking Schools.
Wheelock, A., Bebell, D. J., & Haney, W. (2000a). What can students' drawings tell us about high-stakes testing in Massachusetts? *Teachers College Record,* retrieved March 21, 2001, www.tcrecord.org/Content.asp?ContentID = 10634.
Wheelock, A., Bebell, D. J., & Haney, W. (2000b). Student self-portraits as test-takers: Variations, contextual differences, and assumptions about motivation. *Teachers College Record,* retrieved March 21, 2001, www.tcrecord.org.

CHAPTER 11

No Child Left Behind and Accountability through a Democratic Lens

Cindy Finnell-Gudwien
Williamsville Junior High School, Williamsville, Illinois

> Cindy Finnell-Gudwien, M.A., teaches language arts at Williamsville (IL) Junior High School and is a doctoral student in curriculum and instruction at Illinois State University.

ABSTRACT

> This essay sets forth a theoretical framework for democracy, summarizing five tenets of a democracy. The author then proposes a framework for education in a democracy as suggested by these five tenets. Finally, the author examines the accountability system known as No Child Left Behind through the suggested framework for a democratic education. The author concludes that much of No Child Left Behind does not correspond with the proposed framework for a democratic education.

A Democratic View of No Child Left Behind

The Elementary and Secondary Education Act, or No Child Left Behind (NCLB) as it has come to be known, has been touted by politicians as a demand-

---

An earlier version of this article appeared in *Education and Culture*, volume 21, issue 2, under the title "A Democratic View of 'No Child Left Behind'."

ing means of holding schools accountable for what their students' learn. Its supporters claim it is a fair and reliable method that allows Americans to reasonably evaluate the quality of their children's schools and teachers. Assumedly, this bipartisan law was enacted to ensure an equitable education for all children in our nation. Unfortunately, as countless educators will attest, the many flaws of NCLB cause it to be burdensome and often unfair—anything but the reliable method of accountability it was intended to be. NCLB in its current form does not provide an adequate way to view and judge schools in a democratic system.

For those of us living in America, across ages, races, and geographic areas, there exists an often unspoken concept of what democracy means. From a very young age, our children are taught about the freedoms and responsibilities of our country, the Bill of Rights, the precepts and history of the Constitution. Americans *know* and feel a *sense* of democracy, even if they are often unable to verbalize it in a precise manner. The following is a theoretical framework that attempts to set forth five basic tenets of democracy with which most Americans would agree and then carries over those tenets to the democratic education that is suggested by this framework. Finally, "No Child Left Behind" is examined through the lens of this framework.

What Is Democracy?

Democracy is based on a shared social spirit of mutual interests. A democracy is comprised of people who share mutual interests and a spirit of commitment to these interests. In America, these interests include the well-being, liberty, and equality of all. The common problems, issues, and values of the population are respected, reconstructed, and addressed through the democracy (Hlebowitsh, 1985). For Boyd Bode, this spirit of common interests and purposes was the most important aspect of a democracy (1937).

Democracy is based on a community of cooperation and free interaction between social groups. According to Dewey, a democracy is made up of citizens who participate in full and free interaction between various communities of people. These frequent exchanges in social habit involve readjustment, reorganization, and ultimately progress through the new situations produced as a result of wider relationships and intercourse. A group that isolates itself with the intent of protecting its own interests and keeping its status quo will become static and rigid (Dewey, 1916). Adding to this are James Macdonald and David Purpel who write that in a democratic community, "democracy means participation and community means people in communication and communion" (1987, p. 183). It follows, then, that a democracy requires participatory interaction between groups of people.

Democracy is based on the belief that society will continue to reorganize and progress. Dewey (1916) argued that a democratic society does not rely on a set of customs forced upon its members by a controlling upper class. Rather, it is intentionally progressive, ever widening its interests and encouraging its members to act continuously as guardians over the democracy. According to Henry Giroux, democracy is dynamic by nature and should never be reduced to a set of "inherited principles and institutional arrangements" (1990, p. 364). A democracy assumes that its people will continue over time to provide insight and criticism leading to continued growth and progress.

Democracy is based on respect for individuals. This tenet, of course, is one with which every American is familiar. A democratic society respects its people, despite gender, race, or economic status. Individuality is not only accepted but encouraged as well. As Hlebowitsh (1985) writes, a democracy is socially conscious, respectful of the individual, and it must endorse experiences for both the collective and individual gain of its people. Democracy is based on the belief that all useful services are valued, not only economically favorable ones.

In writing about democracy, Dewey advocated, as most Americans would today, that the members of the community must all contribute to society, that a "social return be demanded from all" (1916, p. 122). Furthermore, he added that this return can be cultural or industrial, an intellectual affair or a social service. In other words, *all* contributions to society, offered at the capacity of the individual, are worthy. In addition, all of the citizens of a democratic society are entitled to the possibilities that would enable them to find their useful roles in society, transcending social class.

What Is a Democratic Education?

The previous section presents a theoretical framework for democracy. This section presents the logical conclusions based on this framework when the five tenets are applied to public education in a democracy.

If democracy is based on a shared social spirit of mutual interests, then education should address the interests of everyone, and it should live in that spirit, not merely prepare for it. In a democratic education, school should not be viewed as merely preparation for future lives, whether for careers or college. School should be viewed *as* life. The activities and learning that students experience in school should serve to construct the mutual interests of the school community, as well as the society as a whole. Macdonald and Purpel write that education should add to that spirit, by freeing students from "barriers to human dignity and potential such as those that come from being poor, frightened, misguided, ignorant, and unaffirmed—in a word, controlled" (1987, p. 187). Indeed, it should bring them to human liberation, a sense of self-fulfillment and freedom. According to Bode, a school based on the philosophy of democracy should be one that students go to learn not only for the future, but to carry on a way of life in the present as well (1937).

If democracy is based on cooperation and free interaction between social groups, then democratic education should allow children to participate and contribute in their education. If, as suggested, democratic education serves in part to help students develop their shared mutual interests, it follows that the pupils should participate and contribute in their own education. The traditional school with teacher as master and students as passive receivers does not exhibit this philosophy. The teacher is not to serve as an external authority, but rather the students should volunteer their own dispositions and interests (Dewey, 1916). Within a school based on the theory of democracy, students should as individuals and in social groups be "alive, active, working hard, inventing, organizing, contributing original ideas, assembling materials, carrying out enterprises" (Rugg & Shumaker, 1969, p. 57). Cooperation, communication, and understanding should exist and enhance the learning and growth. A school's environment and curriculum should reflect the interests of its population (Hlebowitsh, 1985).

If democracy is based on reorganization and progress, then education should be seen as a means of growth for individuals and society. One of the purposes of a democratic education should be to groom citizens capable of questioning and critically examining the basic precepts of society instead of merely accepting and adapting to them (Giroux, 1990). It is this capacity that allows for the growth, reorganization, and ultimately, progress that a true democracy requires. Dewey (1916) wrote that education should acknowledge its social responsibilities and the primary goal should not be teaching students how to make a living, but rather to enlighten and discipline them with the concerns and interests of humanity.

If democracy is based on respect for individuals, then a democratic education should have intellectual opportunities for all, and it should value intellectual variances. A society that claims to be democratic should permit intellectual opportunities to be accessible to all. It is the business of democratic schools to educate fully each individual to his or her capacity. For this to occur, education should allow for and encourage intellectual variances. It should be assumed that all students and citizens will have aims and ideas of their own and not merely be subjected to a few in positions of authority (Dewey, 1916). Furthermore, Dewey added, education cannot be narrowly conceived for utilitarian purposes for the masses and higher education for only a select few. As he wrote: "The notion that the 'essentials' of elementary education are the three R's mechanically treated, is based upon ignorance of the essentials needed for realization of democratic ideals" (1916, p. 192).

If democracy is based on the belief that all useful services are valued, not only economic ones, then education should honor and foster all knowledge. If education based on a democratic philosophy is to offer equal opportunities to all, then it should function under the conviction that all societal services are worthy. The curriculum should operate on the belief that educational activities and experiences should be connected to each other and to the actual lives of the students. Decoding text, for example, cannot be advocated as more socially important than conducting science laboratory experiments or learning about symmetry in an art class because it could possibly lead to a financially better quality of life. In addition, the content should be interrelated and as integrated as possible. As Dewey (1916) wrote, separation and isolation of the content areas leads to a separation of social classes. A democracy cannot use economic return as its criterion for valuing education. The intrinsic social and moral worth should be considered.

Is the No Child Left Behind Act Democratic by Nature?

In January of 2002, President Bush signed into law the reformed Elementary and Secondary Education Act, originally enacted in 1965, now commonly known as No Child Left Behind, 2001. The United States Department of Education's website (www.ed.gov) promoting this law is laced with corporate verbiage, such as "global marketplace," "quality management," "economic leadership," and "financial security." The website proclaims that NCLB will "ensure schools get results" by demanding "more value from the investment." A simple scan of the website shows the law's clear intention—better skilled students lead to better skilled workers and national economic growth. Other than the title, very little of our democracy's education reform agenda will serve to promote the tenets of a true democracy.

NCLB is comprised of four major components: increased accountability, greater flexibility at state and local levels, encouraging proven education methods, and more choices for parents and students. This section will attempt to examine each of these components, comparing them with the ideas of democratic education set forth previously.

INCREASED ACCOUNTABILITY

This law requires schools to measure every public school student's progress in reading and math every year from third through eighth grades and at least once between tenth through twelfth grades. In addition, by 2007, students will also be tested in science. For the first time ever, schools will have their federal funding attached to their standardized tests performance. Schools that show adequate yearly progress (AYP) will be rewarded financially with Academic Achievement Rewards. Schools that do not demonstrate results will have their federal funds reduced. This rewarding and withholding of funds will not eliminate the achievement gap between the rich and poor in our country. Rather, our nation's poorest schools are being penalized financially, thereby rendering them even less likely to succeed.

As mandated by NCLB, each state has set a bar of student achievement that must be met. This first-time bar is based on one of the lowest scoring schools or demographic groups in the state. Every school was to attain this bar after two years and subsequent thresholds every three years until at the end of twelve years all students are to be achieving at proficiency level or higher. Interestingly, the "Testing for Results" section of the NCLB website (ed.gov) states, "if a single annual test were the only device a teacher used to gauge student performance, it would indeed be inadequate." However, an annual test is exactly what the government will use to base its funding decisions.

The website also acknowledges that, "some students score poorly for reasons outside the classroom. A good evaluation system will reflect the diversity of student learning and achievement." Standardized tests alone will neither reflect student diversity of learning and achievement nor consider reasons outside of the classroom for why students might score poorly. The site further admits that testing "does sometimes cause anxiety . . . young people need to be equipped to deal with it." Despite accounts of young elementary students crying because they are nervous about the high-stakes test about to be undertaken, no cure is offered to help them or their teachers "deal with it."

Is the call for increased accountability democratic? It does not seem to be. Testing done in up to three content areas almost yearly does not address the interests of everyone and definitely views education as mere skill preparation for a future career. Mandates handed down from upper-level politicians with no dialogue with the vast majority of children, teachers, and families who are affected by them are not based on a democratic spirit of cooperation and interaction. Using one standardized test each year to determine math and reading skills does not hold schools accountable for what a truly democratic education should be. Nowhere in this system are critical inquiry and human enlightenment assessed.

To assume that students are learning to become citizens capable of realizing the democratic ideal because they are sufficient in drill exercises in reading and mathematics is woefully misguided. Furthermore, the call for higher accountability via standardized testing is undemocratic because it does not honor all knowledge. It plays the power card, deciding for Americans what makes one legitimately educated. It comes close to the ideological attack that Michael Apple (1993) writes is so dangerous, especially when done on the cheap.

GREATER FLEXIBILITY FOR STATE AND LOCAL LEVELS

The second major component of NCLB is a claim for greater freedom at the state and local levels. States were permitted to create, and for the most part pay for, their own assessments. Local schools were given the freedom to make spending decisions with up to fifty percent of the non-Title 1 federal funds they receive. Funds, therefore, could be transferred from one account to another. Schools that decide to teach to the test or already have demographics that will probably result in proficient scores, at least in the first few years, will be financially rewarded. Schools with low scores, the very schools in need of added financial assistance, will be penalized.

Meanwhile, most schools are relying heavily on property taxes for their school funds to begin with, already an undemocratic process. Neighborhoods with many businesses and high-priced homes send more money to their schools, while struggling, depressed communities have very little to contribute to their local schools by way of property taxes. Students in these poor areas usually receive the state minimum per pupil. And lest we forget, public school systems are already losing billions of dollars annually to corporations given property tax breaks (Weaver, 2003). With the already inequitable state of affairs of school funding, NCLB dishonors the system further in tying funding to testing for the first time ever, further setting up our most economically deprived schools for failure.

If NCLB truly granted greater financial freedom for the state and local level, then perhaps it would be democratic. Schools would be able to address all of their population's interests and funding could be progressive in nature. In reality, however, school funding could not be less democratic. The amount of money spent per pupil varies by thousands of dollars. Students in poor schools have less qualified teachers, fewer resources for learning, and less community support, all of which contribute to lower test scores. Under this system, individuals are not respected, and the playing field is not level, yet NCLB acts as though it were.

ENCOURAGING PROVEN EDUCATION METHODS

This component can be broken down into two parts: teachers and methods. To begin with, NCLB states that all teachers were to be highly qualified by the

2005–2006 school year. They must hold at least a bachelor's degree, have full state certification, and have demonstrated competency in their subject area. To do this, NCLB allows states the flexibility to find alternative certification routes, often creating "short cut" approaches to becoming a teacher, reducing or eliminating the education coursework. At the same time, it ignores the unique circumstances of rural, inner-urban, and middle schools, which all frequently employ teachers trained in education to teach in different content areas.

The law also gives districts the right to reward good teachers with merit pay and to give bonuses to teachers in high-need subject areas, such as math and science. Districts will be "free to use their funds" for these various suggestions, according to the United States Department of Education's website. With schools in high-poverty areas unable to pay teachers competitively to begin with, NCLB permitting them to pay teachers more does not matter.

In addition, NCLB states that one percent of the funding for this program is set aside for the Secretary of Education to award grants to states that assess their teachers' performances using gains in student academic achievement. In other words, states willing to assess their teachers based on their students' performances will be rewarded financially for doing so. This will not encourage quality teaching but rather teaching to the test.

The second aspect of this component is the call for proven methods. Only scientifically proven approaches and programs in reading and mathematics, and later, other areas, will be funded. Education programs, or "fads," as the ed.gov website refers to them, that have not undergone rigorous scientific research will not be supported. Instead, schools that use officially tested and empirically sound programs will be financially rewarded. To decide which programs constitute proven methods and which are "fads," the United States Department of Education has established the What Works Clearinghouse, an organization developed to identify approaches in education that have been scientifically proven to be successful. Their website states, "Over time . . . parents will be able to ask their principal, teachers and school board members about the extent to which they select programs and curricula that the research has determined to be effective." Teachers will be expected to know what is on the What Works Clearinghouse's approved list of approaches and to teach using only those approaches.

In addition, the What Works Clearinghouse website (http://ed.gov/nclb/methods/whatworks/whatworks.doc) states that teachers "need a central, trusted, and independent source of scientific evidence of what really works in education." It declares, "conflicting interpretations and disagreements about a study's finding too often cause confusion among education practitioners." Statements such as these convey the idea that teachers have no informed knowledge about how and what to teach their students. Their educational experience, knowledge in the field, and knowledge of individual students don't provide enough information for teachers to know what really works in their classrooms, according to NCLB.

While providing information on the reliability of educational methods to educators and the public is a good practice, attaching school funds to only those programs that make the Clearinghouse's list is not. How can NCLB promote democracy when it stifles individuality and progress by forcing teachers into certain methods and tying funding to them?

MORE CHOICES FOR STUDENTS AND PARENTS

If, under NCLB, a school is labeled as failing due to low standardized test scores, the parents of children in that school have the option of transferring their child to another higher-scoring public or private school. This will be at the expense of the failing school. In addition, failing schools may be required to provide supplemental services to their students, again at their own expense. These services could be tutoring and mentoring, after-school programs, or remedial classes. The "Facts About Faith-Based Efforts" portion of the Department of Education website suggests that faith-based organizations can provide these programs.

When the website shifts to this fact sheet, the rhetoric changes considerably. The corporate speech is replaced with loaded phrases such as "spread the message" and "rally the armies of compassion." This is clearly a call for private organizations to improve or replace, if necessary, public education. It is undemocratic to arrange for weak, penalized schools to fail. It is unlikely that requiring a failing school to use some of its precious funds to send its pupils to other schools will enable it to crawl out of the failing category.

The transferring of students from failing schools to successful ones is being handled undemocratically. Students in schools not reaching state AYP often have populations including families with limited English, limited literacy, limited resources, and limited leisure time. These parents are unlikely to take advantage of their right to transfer their children to better schools. In Springfield, Illinois, for example, six elementary schools were labeled as needing improvement in 2003, making 148 children eligible to transfer for the 2003–2004 school year. Two did. This reflects the trend across the state and nation (Friedman, 2003).

In addition, this component is highly under-funded. After the 2003 school year, Chicago Public Schools had 133,000 students eligible for reading and math tutoring under NCLB rules, but they could afford to provide tutoring to only 25,000 to 30,000. Similarly, thousands of children in Chicago who were eligible to transfer under this reform were not provided the opportunity due to lack of space in better schools. In Chicago, 270,000 students were eligible for "choice" of schools; 19,000 applied; there were a mere 1,097 spots available; and at last count, about 500 actually transferred (Rado & Olszewski, 2003).

Conclusions

No Child Left Behind is not an adequate method of making schools accountable, and it does not meet any of the previously set forth characteristics of a democratic education: a) it focuses solely on pre-determined skills to prepare students for a financially secure future, b) it lacks free interaction and participation, c) it does not promote societal progress, d) it lacks respect for individual variances, and e) it does not value all knowledge.

Never does it acknowledge factors other than teachers and methods that are critical to the success of a school: class size, building facilities, demographics, family support, socioeconomic status, mentoring for new teachers, amount spent per pupil, and administration, to name but some. Yet schools and their teachers are the ones that are held accountable for all of these factors. Our legislators represent the interests of their constituents, and therefore, in theory, NCLB represents the will of the people. In reality, however, this law does little to support the people of our nation and is, therefore, not democratic and certainly not a valid measure of schools' performances.

This year, millions of children across our nation will walk into their classrooms, sharpen their no. 2 pencils, take their seats, and begin to take high-stakes standardized tests. Some of them will have been beaten that morning, others will not have had breakfast or even dinner the night before; some will have colds, headaches, or other health problems; some will be grieving over recent deaths in their families; some will be worried about typical adolescent problems; some will be overly anxious about taking the test; and some will simply not care about the test. All, however, will take it. All will be part of this widespread instrument of accountability, because, as our nation's president and Congress have declared, *no* child will be left behind.

References

Apple, M. W. (1993). The politics of official knowledge: Does a national curriculum make sense? *Teachers College Record, 9.5*(2), 222–241.
Bode, B. H. (1937). *Democracy as a way of life.* New York: The Macmillan Company.
Dewey, J. (1916). *Democracy and education: An introduction to the philosophy of education.* New York: The Free Press.

Ed.gov. *U.S. Department of Education: Promoting Educational Excellence for All Americans.* Retrieved April 26, 2006, from http://ed.gov.
Friedman, L. (2003, September 22). Two students will transfer. *State Journal-Register* (electronic version).
Giroux, H. A. (1990). Curriculum theory, textual authority, and the role of teachers as public intellectuals. *Journal of Curriculum and Supervision, 5*(4), 361–383.
Hlebowitsh, P. S. (1985). Democracy and education. *Illinois Schools Journal, 65,* 33–39.
Macdonald, J. B., & Purpel, D. E. (1987). Curriculum and planning: Visions and metaphors. *Journal of Curriculum and Supervision, 2*(2), 178–192.
Rado, D., & Olszewski, L. (2003, September 13). Most in failing schools will be denied tutoring. *Chicago Tribune* (electronic version).
Rugg, H., & Shumaker, A. (1969). *The child-centered school.* New York: Arno Press & *New York Times.*
Weaver, R. (2003). Keynote as Prepared for Reg Weaver, President National Education Association to 2003 NEA Representative Assembly in New Orleans, LA. Retrieved September 14, 2003, from www.nea.org/speeches/sp030703.html.
What Works Clearinghouse. (n.d.). Retrieved April 26, 2006, from http://ed.gov/nclb/methods/whatworks/whatworks.doc.

Summary and Implications

Michele Kahn
University of Houston–Clear Lake

Mimi Miyoung Lee
University of Houston

Carrie Markello
University of Houston

Heidi C. Mullins
Eastern Washington University

Annapurna Ganesh
University of Houston

The common theme spanning chapters 9 through 11 is the feeling of marginalization that pervades teachers and teacher educators when they are forced to redefine the meaning of education in terms of standardized test scores. Given the noble goal of the No Child Left Behind (NCLB) Act, which is to ensure equitable education for all children in our nation, as described by Finnell-Gudwien, it is ironic that educators see NCLB as the main hindrance of implementing both an equitable and quality education in their respective settings. The experiences and findings of researchers and teachers affected by the NCLB mandate are discussed in these three chapters.

Kosnik dealt explicitly with the culture shock and reclamation as a teacher educator entering the U.S. educational policy environment, which is heavily regulated and punishes low-performers. Through his interpretation of three teachers' drawings, T. Ganesh critically examined teachers' repressed opinions concerning high-stakes testing through the use of visual representations. These teachers felt bound by NCLB and the testing movement. As a result, they experienced limitations in their creativity and ability to teach. Finnell-Gudwein discussed notions of democracy as they explicitly relate to NCLB, revealing incongruent goals of equity and accountability in achievement testing.

Despite advances in cognitive psychology and its contributions to learning and motivation theories, educational policy continues to uphold classic behaviorist notions of schooling. The most basic form of behaviorist theory assumes that learning is nothing more than acquiring new behavior. In this time of behaviorist policy, educators are placed in a mechanistic situation. For example, Pavlov's experiments in training dogs to salivate on command with certain stimuli later became known as "conditioned reflex" which evolved into the hallmark of behaviorist psychology. In this sense, accountability, in the form of testing and standards, seems to take the role of stimuli. However, just as classic behaviorist techniques function extrinsically, educational policy's restrictive impositions ensure that accountability measures falter because they do not attend to the human qualities and inner lives of students, teachers, and teacher educators.

These personal and collective narratives of how the current format of standardized tests—by which the quality of teaching is measured and the destiny of schools is determined—demonstrate how education becomes dehumanized, thus moving further away from democratic ideals. Education is more than what can be shown solely by test scores. In this regard, it is no coincidence that all three studies emphasize the importance of qualitative research methods and alternative forms of assessment and representation. The authors in this section use visual data, interviews, self-reflection, and interpretive analysis to reveal layers of reality that otherwise would be shadowed and silenced by the current preoccupation with a narrow version of science-based research. These chapters remind readers that the real meaning of education should be found in the very "process" of the learning experience, not the measurable "product" of one right answer. They also suggest that educational means and ends need to dialectically inform one another.

Implications drawn from the three chapters include the importance and impact of context-driven work, the value of mixed method approaches to research and evaluation, and the need to strike a balance between individuality and consistency in teacher education programs. Additionally, these authors, each in their individual terms, spotlight factors that demand consideration for educating our nation's future generations. These include, but are not limited to: the political agendas that marginalize teacher education research; the visceral messages that invoke a sense of alienation, hopelessness, and lack of control; the de-valuing of teachers' knowledge regarding effective practices in the face of high-stakes testing pressures; and the current reality that offers little in terms of substantive democratic education.

Division 4
UNDERPINNINGS OF POWERFUL ACCOUNTABILITY SYSTEMS

Overview and Framework

Neil J. Liss
Willamette University

> Neil J. Liss, Ed.D., is a visiting professor of education at Willamette University in Salem, Oregon. His scholarship focuses on philosophical and spiritual issues in education and on the aesthetics of qualitative research.

The final three chapters in this edited volume explore how certain forms of accountability can provide a richer construction of the educational experience. They contend, in part, that measured evidence of learning occurs as the result *of* these experiences. Understanding the "underpinnings," examined herein, should help focus educators on the relationships crucial to powerful learning and on documenting the collective processes needed to make sense of these experiences.

In chapter 12, The Antecedents of Success: The Finnish Miracle of PISA, Estola, Lauriala, Nissilä, and Syrjälä provide a comprehensive explanation of how the socio-cultural context of Finland combines with the dialogical practices of pre- and in-service teachers. This nexus positions accountability differently than traditional models of assessment. The educational endeavor must recognize schooling as integral to the social system and must incorporate the teacher as an autonomous agent in this process of recognition.

In chapter 13, Creating a Learning Space for Educators: Policy Development for Accountability Systems, Kwo links a similar understanding of education's place in the larger picture of society. Dedicating "supportive" learning spaces for teachers to reflect on that placement is a necessary part of that linkage. She holds out the effective role of "mutuality" for both leadership and pedagogy to support each other's growth and accommodation to the increasingly pressurized expectations of external educational stakeholders.

And finally, in chapter 14, Lessons from Te Kotahitanga for Teacher Education, Bishop offers how endemic "challenges" to the learning of marginalized populations can be met through directed confrontation with the discursive practices that shape social consciousnesses. Since knowledge created within the educational setting privileges one set of meaning systems, every educator must reflect upon their role in normalizing these practices. Understanding is crucial; data of all kinds need to be available in order to make the most holistic and accountable forms of meaning. Implicitly, and sometimes explicitly, each of these chapters directs us to think through how contiguous knowledge and place are; what are some ways to generate this understanding; and what do these ways mean in terms of forms of accountability.

Framework

Two broad terms frame this section: action research and transparency. Both are increasingly ever-present in the role call of research journals; both inherit the baggage of justifying certain educational practices in an era of intensifying standardization.

Action research, for starters, comes not merely as a reaction to hegemonic forms of accountability, codified in the 2001 No Child Left Behind law that nationalized standardized testing (see www.whitehouse.gov/news/reports/no-child-left-behind.html). There are enough responses that reify the law's "gold standard" of randomized selection research (particularly the "What Works" Clearinghouse at www.whatworks.ed.gov/ and the Campbell Collaboration at www.campbellcollaboration.org/). Action research, particularly the kinds articulated in this section, exposes the flaws in this scope of scientism. Embedded within narrative inquiry (see Connelly and Clandinin, 1990; Craig, 2003), care theory (especially Noddings, 1992), and critical pedagogy (Apple, 1999; McLaren, 2002; Kincheloe, 2004), this fluid and flexible form of inquiry (Schwab, 1982) brings awareness to all expressions of knowing that place human experience transcendent to, though dependent upon, the sense made out of them.

Action research, herein, broadens into the various types of knowledge that come from participation in these human experiences, whether in our schools or elsewhere. These include efforts in teacher research of personal practical knowledge and case study research, participant narratives, and student-generated narratives. Spiritual and meditative, qualitative and poetic, empirical and documental, all these forms push research toward holistic acceptance of the full range of human experiences that occur in the process of trying to educate. Otherwise silenced in the privileged forms of scientific accountability, these forms include the voices of students both "doing and undergoing" learning (Dewey, 1934/1958); the agency of teachers coming to understand the paradoxical and diffusive complexity of what powerful teaching and learning means under whatever construct shapes their practice; the growth of school leaders in their struggled negotiating of the "swampy ground" (Schön, 1983) of lived exigency in their schools; and, even more often forgotten, the rest of society, for whom their local school roots the basic tether to its morals and values.

Moreover, action research offers a more dynamic and accessible way to account for the multiple crossroads that 21st-century education must traverse. In these encounters, education needs both stable and fluid inquiries, due to the nature and complexities of human beings in comparison to the innumerable and inanimate variables thrown upon us. Stable forms allow a starting point to access these complexities, while the flexible forms of knowing permit growth and development not accessible to stable conceptions of human experience. A partial list of these issues may include questions of human authenticity and autonomy; human/environment interactions and their ecological outcomes; the moral and spiritual values that ground communal understandings; and the constant flow of change that engulfs every society despite their attempts at stabilization and reform.

All these concerns become the seed of action inquiry, as educators help position their students within an ethical relationship to these questions. Thus, a "bigger" perspective on accountability is needed to provide a start for growing a commitment to think through and confront constantly emergent exigencies. Whole editions of professional and scholarly journals have addressed this (see *Phi Delta Kappan*, v. 87, no. 3, November, 2005). Diverse disciplines have also generated research, such as art education (see Tavin & Hausman, 2004) and environmental science education (Payne, 2006). These are but a, yes, random sampling of the large-scale commitment to the forms of knowing that hold *accountability* accountable to something more than outcome-based schooling.

But the following chapters also give life to the inescapable question of whether we can trust that what calls itself "accountability" really is. Put differently, these chapters ask: What is the quality of what we are presented with as evidence? Is the form of accountability a diversion into stagecraft, a façade germinated from prior and punitive expectations? How much of that quality is lost when placed into systems of knowing that spur replication and systemizing? How is what we are told being accounted for? Is the learning experience itself "accountable"? Or rather, how does any form of accountability relate to, be accountable to, our reasons for educating? The human stories, the human face of that aspiration, must be present and knowable for true accountability to underpin successful education.

This division of the handbook, "Underpinnings of Powerful Accountability Systems," provides three ways of addressing these quandaries. Together, they offer possible, hopeful, and realistic pathways for reflective, collaborative inquiry and the constant critical distancing necessary to understand every attempt at understanding. These are needed to connect teachers to students and students to community in ways that create the conditions for the kinds of accepted and anticipated accountability we have become inured to from our schools.

References

Apple, M. (1999). *Official knowledge*. New York: Routledge.
Connelly, M., & Clandinin, J. (1990). Stories of experience and narrative inquiry. *Educational Researcher, 19*(5), 2–14.
Craig, C. (2003). *Narrative inquiries of school reform: storied lives, storied landscapes, storied metaphors*. Greenwich, CT: Information Age.
Dewey, J. (1934/1958). *Art as experience*. New York: Capricorn Books.
Kincheloe, J. (2004). *Critical pedagogy primer*. New York: Peter Lang.
McLaren, P. (2002). *Life in schools*. Boston: Allyn and Bacon.
Noddings, N. (1992). *The challenge to care in schools*. New York: Teachers College Press.
Payne, P. (2006). Environmental education and curriculum theory. *Journal of Environmental Education, 37*(2), 25–35.
Phi Delta Kappan, 87(3), November, 2005.
Schön, D. (1983). *The reflective practitioner: how professionals think in action*. New York: Basic Books.
Schwab, J. J. (1982). The Practical: a language for curriculum. In *Science, curriculum and liberal education*. Chicago: University of Chicago Press.
Tavin, K., & Hausman, J. (2004). Art education and visual culture in the age of globalization. *Art Education, 57*(5), 47–52.

CHAPTER 12

The Antecedents of Success
THE FINNISH MIRACLE OF PISA

Eila Estola
Oulu University, Finland

Anneli Lauriala
University of Lapland, Finland

Säde-Pirkko Nissilä
Oulu University of Applied Sciences, Finland

Leena Syrjälä
University of Oulu, Finland

> Eila Estola, D.Ed.Sci., is a senior researcher with a background in early childhood education in the Faculty of Education at Oulu University. Her main research interests are the moral dimensions of teaching and teacher identity.
>
> Anneli Lauriala, D.Ed.Sci., is a professor of education at the University of Lapland. Her research inquiries center on teachers' professional development and school and teaching innovations.
>
> Säde-Pirkko Nissilä, Ph.D., is a principal lecturer in the School of Vocational Teacher Education, Oulu University of Applied Sciences. Her research interests are teachers' professional development, individual and collective reflection, and interactions in educational contexts.
>
> Leena Syrjälä, D.Ed.Sci., is a professor of education at Oulu University. Her research interests include school reform, evaluation, and the contradictions of teaching.

ABSTRACT

The Finnish learning context, along with Finnish teachers' perceptions of their teaching experience and teacher education, are discussed as factors explaining Finland's PISA results. The country's

Nordic welfare ideology and sense of social justice are also brought into view. An overview of teacher education and examples of action research and inquiry-oriented teachers' research movement are additionally discussed. The chapter concludes by arguing that the pedagogical autonomy of Finnish teachers is relatively high and is a critical factor in students' achievement results. At the same time, a shift toward managerialism, privatization and accountability is underway, which, if educationally dysfunctional, could lead to a decrease in teachers' professional autonomy and negatively impact students' learning achievement.

Introduction

PISA (Program for International Students Assessment) shone the spotlight on Finland as a model country of basic education. The results of the first year (Organization for Economic Co-operation and Development [OECD] 2001), which showed Finnish adolescents to be the best readers in the world, made other countries envious of Finland's PISA miracle and interested in Finnish schools.

Finland similarly produced exemplary results in the 2003 PISA program. In that year, Finnish adolescents showed greater mathematical proficiency compared to the other OECD countries and continued to hold the number one position with respect to reading skills. Finns also ranked high in natural sciences and showed excellent problem-solving skills. Overall, the PISA results indicated that Finnish adolescents achieve a high standard of learning and that there were no major regional differences in the country. These findings underscore the high standard of Finnish basic education and the equal opportunities for learning that exist (Kupari et al., 2004). By way of contrast, results obtained by the United States were below the OECD average in all areas except reading skills, which were at the OECD average.

The success story of Finnish schools has given rise to an active political discussion. Researchers have thrashed out the causes of this success. The topic has been discussed in thematic issues of various journals, including *Zetshrift für Pedagogik* (2003, vol. 49) and the *Scandinavian Journal of Educational Research* (2004, vol. 48).

Finland's success in the 2003 program was attributed to a number of factors. Education is highly valued in Finnish society. Equal opportunities for basic education, regardless of social status, sex, or ethnic background, have been praised as strengths in the Finnish educational system. Also, funding for basic education is the responsibility of the municipalities. Hence, schools are located close to students and their homes. Citizens trust in education, and a long tradition of cooperation between schools, public authorities, and homes exists.

The attention given to teacher education in Finland additionally guarantees high-quality public school education. Simola (2005) underlines the significance of the aforementioned factors related to the Finnish PISA miracle. While he does not wish to underestimate the pedagogic background of high quality teacher education programs and excellent teachers in the PISA success, he prefers to attribute the success story of Finnish schools to social, cultural, and historical factors, which need to be recognized in influencing the PISA success. Siljander (2006) similarly highlights the development of the Finnish welfare state and its significance as the background of the PISA results.

The Nordic Social Welfare Model as a Context of Education

Finland is located in Northern Europe and borders Sweden, Norway, Estonia, and Russia. Although the country covers 338,000 square kilometers, Finland has only 5.2 million inhabitants with a rather homogeneous population. When the criterion of gross domestic production per capita income is applied, Finland is one of the 20 richest countries in the world and enjoys small differences between the rich and poor, although a gap is beginning to develop.

Finland and other Nordic countries embrace a form of social justice that supports the welfare of all inhabitants; Finnish society tends to care for everyone, especially those who are helpless, like children, old people, sick, handicapped, and other vulnerable individuals. Many social and health services as well as education and schooling are free and available to everyone. To finance these services, Finns pay relatively high income taxes (Moore, Antikainen, & Kosonen 2005, p. 26–27). Through its educational system, Finland aims to minimize the marginalization of young people. Special attention is also paid to pupils with learning difficulties to ensure that all pupils attain at least the basic goals of education.

A discussion of the quality of Finnish education would not be complete without looking at the social context, including Finnish family policy and early childhood education. The objective of Finland's family policy is to create a safe growth environment for children and to guarantee parents the material and psychological resources to bear and raise children. It covers the costs of childcare so as to prevent children from becoming an excessive financial burden on their families.

Care for children begins before a baby is born and continues in the form of child guidance and family counseling for every mother and father. The child benefit is paid from State funds for the maintenance of every child under 17 years of age. The total length of maternity and parental allowance leave is 263 working days. Parents with children under three years old have the right to take unpaid childcare leave from their work.

The system is well-developed, stable, and highly appreciated by parents. It is characterized by sensitivity to the rights of children, and concern for equality and fairness is avowed throughout the system. The Ministry of Social and Health Affairs has a central responsibility for the education and care of children under six years old. The payments for daycare depend on the incomes of the families but are reasonable. Free pre-school education for six-year-olds has been provided since 2000 (OECD 2001, p. 163). Early childhood education is based on the educare concept, in which care, education, and instruction are combined. In Finland, educare is predominantly provided by municipalities (Kess, 2002; Välimäki, 1998).

Pre-school constitutes an important link between daycare and primary school education. Teaching methods are drawn from early childhood education with play and peer group activities. The core curriculum is based on the content areas but not on specific subjects. Readiness and willingness to read, to perform mathematical tasks, and to engage in the arts are supported and, more importantly, academic skills are not considered more valuable than play.

Equal access to education underpins the Nordic welfare society. To support equal opportunities to learn, Finland has built a comprehensive basic school system free of charge. The objective of basic education is to support pupils' growth toward humanly and ethically responsible membership in society and to provide them with the knowledge needed in life. Finland's compulsory school age is from seven to 16 years old, which translates to nine years of basic education. Also, the necessary schoolbooks, other study materials, and work equipment are provided to pupils free of charge. A very special feature of Finnish schools is a healthy and supervised hot meal, which is available free of charge on every school day to every student (Linnakylä, 2004, p. 38; OECD, 2000, p. 5).

Although Finland is known as a country of equality in education, it is important to take a closer look at the PISA results and the background factors. Research (OECD, 2004a) has shown that students' attitudes, self-efficacy, interest, and motivation are more significant for learning than their home backgrounds. Parents' interest in and support of school work are, nonetheless, vital as well as the support of the whole living environment. Thus, the differences in family backgrounds such as socio-economical status, parents' professional status, and cultural environments have some influence on the results: some groups of students do better in school than others.

In Finland, the socio-economic level of parents has a weaker influence on the results in mathematics than in other European countries on average (OECD, 2004a). These results challenge earlier findings that show that the socio-economical background of homes is related to learning outcomes in mathematics (Beaton & O'Dwyer, 2002; Yang, 2003). Furthermore, the significance of socio-economic status was even narrower in reading ability than mathematics.

The educational status of parents has proven to be a central factor in the PISA results. Finland had the highest proportion of educated mothers in all of the participating countries (58%). The children whose mothers had a tertiary or secondary education did better in mathematics than the pupils whose mothers had a basic education. Although the statistics show that the smallest differences of results between the children of tertiary and basic education mothers were in Finland, Iceland, the Netherlands, and Australia, Finland still faces the challenge of offering equal educational opportunities to everybody (OECD, 2004a, 2004b).

The cultural environments of home, school, and society are closely connected with school achievement. Research shows that the cultural level (evaluated according to the amount of classical literature, poetry collections, products of arts, etc. in homes), especially in high socio-economic households, was above the average of the other OECD countries. Even the lowest group of homes was above average. This was connected to the reading results especially, and also to mathematics achievement, which suggests that the cultural stimuli and the value given to culture are generally reflected in the learning results (Linnakylä, 2004). In general, Finnish literature is popular and ample, which cultivates a national stance of positive values and attitudes toward reading. The country's literacy level in Finnish is nearly 100%.

In discussing Finnish teachers' experiences and teacher education as factors explaining the good PISA results, the authors assert that teachers are critically important to Finnish adolescents' abilities to attain the highest results in the PISA program. We have all worked for over 20 years as teacher educators and conducted research on teacher development, teacher thinking, and teacher knowledge. We all share an interest in teachers as independent pedagogic professionals who know, hope, and care. We approach the PISA results from the perspective of our respective research orientations and the national discussion on the topic, which encourages teachers and student teachers to express their opinions.

Teacher Education

An important explanation for the fine PISA results is undoubtedly the high-quality teacher education and excellent teachers in Finland (Simola, 2005, p. 455; Linnakylä, 2004, p. 39). All Finnish teachers, from early childhood education to vocational and adult education, have university degrees. Teachers in early childhood education have either bachelor degrees from universities, or they have graduated from universities of applied sciences (polytechnics). Also, teachers in elementary, secondary, or senior high schools, as well as secondary and tertiary vocational teachers, have the minimum of a master's degree. Teacher education is compulsory for all teachers and is pursued either during master's degree studies or afterwards.

Teaching as a profession is very popular in Finland. For instance, only approximately 10% of all applicants are admitted to teacher education programs yearly. Besides good grades in school and high scores on university entrance exams, students are interviewed to assess their motivation and aptitude toward teachers' work. This multi-phased selection process has led to student candidates who are highly motivated and multi-talented in academic studies, music, arts, and sports (Linnakylä, 2004, p. 39).

In early childhood and teacher education, the vast majority of students are women. Traditionally, teaching has been considered worthwhile work for females. The relatively high proportion of females in the teaching profession has not changed much over the years, despite the fact that teaching children has gained professional status. Teaching attracts males, although they are somewhat poorly paid compared to other, predominantly male professions. Subject area teacher education attracts both males and females, and, in vocational teacher education, the number of men and women is equal.

From our point of view, being a professional teacher means attaining research-based orientation to one's work, which involves a moral commitment toward the "other," the child, the youngster, the student (Estola, 2003). Without this sense of moral responsibility, it might be hard, or even impossible, to commit to work with children and young people (Hansen, 1998). Consequently, teaching as a moral undertaking is a starting point in all Finnish teacher education programs. Ethics courses are included in the programs, and many student teachers talk about their moral commitments when reflecting on their motives and attitudes toward the teaching profession (Niemi, 1988).

Paying attention to the intertwined personal and professional commitments with emphasis on the moral basis of teaching suggests Finland's preference for teacher education rather than teacher training. Moral commitment is developed throughout the teacher education continuum from early childhood to vocational and secondary teachers. Vocational student teachers, for example, state: "Teaching is an ethical profession" and "Democracy and justice are the cornerstones" (Nissilä, 2005a, p. 214). Ethical consciousness and a serious desire to provide the best for Finnish children prompted another student teacher to state:

> I want to recognize and overcome the problems of being a student, child, person. This sounds wonderful on paper, which is why I want to see myself to achieve this goal over and over again, time after time, tirelessly and with tenacity. . . . I really hope that this were not mere prose and an overused cliché. (Estola, 2003, p. 192)

One reason for Finnish student teachers' high commitment to children and young people may be the multi-phased teaching practicum, which starts with a course in educational psychology and involves classroom observations, discussions, and reports based on theory and empirical data gleaned from observed actions, interactions, and developmental assessments. The relatively long practicum is counseled by different types of supervisors, mentors, and tutor teachers from both the university and schools. A central aim is to link theory and practice.

As stated above, the central mission of teacher education in Finland for over 20 years has been research-based professionalism (Westbury, Hansen, Kansanen, & Björkvist, 2005). A thesis is a requirement of every teacher education program in Finland. In the 1970s, when teacher education became more academic, changes were made to its course contents. A considerable amount of research was added. Elementary teacher education became a four-year program leading to a master's degree. In secondary teacher education, the aim was to promote "a teacher as a researcher of his own work."

While working on their research papers, student teachers begin to see the close connection between research and the practical act of teaching. They, together with their teacher educators, often use action research approaches in developing their projects. For instance, action research was used in the Department of Teacher Education at Oulu University to develop field-based reflective teacher education, ethics education, and multicultural teacher education programs (Räsänen, 2002; Nissilä, 1997; Syrjälä, 1998).

In general, action research has been a widely accepted approach in Finnish teacher education, and the inquiry-oriented teachers' research movement has attracted several hundred teacher participants (Korpinen, 1996). Research is the underlying principle permeating most courses, and such research highlights self-learning and learning communities while enhancing teachers' professional judgment.

In seminars, student teachers develop and discuss their research ideas, usually pedagogically-oriented, or related to practice or their personal experiences in school. Some student teachers who participate in small-scale development projects find this opportunity personally significant. A study on undergraduate students' experiences of research on school innovation (Lauriala & Syrjälä, 1995) came to the same conclusion. The group included both student teachers and experienced teachers. The seminar sessions focused upon the development of the participants' collaborative skills as well as connecting practical experiences to theoretical studies. Consequently, these interactions modified the participants' perspectives of school, learning, children, and research.

The participants found this activity highly significant to their professional growth and especially appreciated collaboration as reflection that was carried out jointly. Collaboration with university researchers was considered essential, because a teacher cannot always be a devoted researcher and a teacher (Lauriala & Syrjälä, 1995).

The participants of the regular teacher education program felt that the research project made them more confident about their ability to develop their work:

> My own research was closely connected with teaching. A teacher as a researcher provides a new viewpoint to teaching. The class I am teaching is very suitable for various experiments. I hope to continue inquiry in the future, too. (Lauriala & Syrjälä, 1995, p. 110)

Reflection is aimed at helping student teachers to develop multiple goals and encourage transformative development. Transformation is not one significant emotional event; rather, it is a series of experiences that teach student teachers to think from a critical perspective. Reciprocal processes enable the students and teachers to construct meanings. They occur within the context of relationships. It concerns, for instance, group reflection in seminars focusing on research and practical teaching among primary, secondary, and vocational student teachers as well (Nissilä, 2006). One student teacher wrote the following about her seminar group:

> I realized the importance of scientific study . . . and that practical applications and research work are not two separate worlds but they should have dialogue in the form of practice. (Nissilä, 2006)

In innovative teacher education projects, reforming or transforming teacher education spurs reform in schools, especially in those where student teachers practice. This also means that the pupils/students must become researchers of their learning. This is a good starting point for encouraging student teachers to engage in critical analysis and change their taken-for-granted views and practices (Lauriala, 1995; Lauriala, 1997; Lauriala & Syrjälä, 1995).

For instance, the 1980s ground-breaking projects—with their practicums in innovative, learner-centered classrooms, in teacher education in Oulu—seemed to have conveyed to most of the participants a view that the curriculum is not to be taken as given but that teachers have freedom to select and organize the content. This new perspective meant that they no longer adhered to the curriculum blindly, but modified, selected, and applied it to their own needs and interests. The participants of these courses and many of those who were influenced by them in nation-wide in-service courses began to see teachers as interpreters of curriculum and stressed the necessity of adapting the curriculum to the needs of individual students (Lauriala, 1997, p. 85; Lauriala, 1992, p. 529). Interpreting curriculum is a necessary condition for producing independent learners and thinkers; such learning and thinking cannot be achieved by teaching students a prescribed curriculum (Lauriala, 1997, p. 86).

Particular participatory and collaborative research, endeavors where teachers, researchers, and policy-makers or administrators work together, has increased in our country and has been able to change the status quo. Thus, the pupil-centered, informal primary school pedagogy has successfully prepared the ground for teaching in the secondary schools, this being reflected in excellent PISA results. The development has taken or is taking place particularly through the increase of student-centered approaches. Some of the innovations have been teacher-initiated grass-root enterprises, which have spread to others through grassroots programs based on work in innovative classrooms and through peer learning (Lauriala, 1995, 1997). The importance of continuing development projects and life-long learning has been realized in all forms of teacher education and in-service education in Finland (Nissilä, 1997, p. 70).

Teacher Status, Autonomy, and Restructuring of Education in Finland

Finns traditionally value teachers and school. After World War II, a school house was built in every village. Teaching was a popular occupation in the post-war period, because it was the easiest route for rural young people to improve their social status. Finnish teachers typically identify with the upper social strata in society, and their political opinions have been rather conservative, but the situation is changing (Simola, 2005).

Teaching as a profession is still highly appreciated and has remained as one of the most popular higher education options. Over the past few decades, however, changes have taken place in the regulation and structuring of the teaching profession in Finland. After the adoption of the comprehensive school system in the 1970s, there was a shift of focus from the policy of equality and centralized decision-making and planning toward deregulation, decentralization, privatization, and goal attainment (Lindblad, Johannesson, & Simola, 2002). This shift is reflected in everyday teaching and is particularly apparent in teachers' attitudes toward school reforms, collaborative undertakings, and continuous curricular development.

The implementation of comprehensive school reform in the 1970s was a fully centralized and politico-administrative process, whereas the curricular development since the end of the 1980s has been rhetorically marketed as a teacher-managed development. The earlier development seemed to parallel what was taking place in the other Nordic countries as well as Great Britain with a national curriculum, teacher evaluations, and accountability for the outcomes. According to Goodson and Numan (2002), this resulted in the underestimation of the local tacit knowledge of the school communities and an underrating of the teacher's role. This devaluing of the school and teachers, we submit, can be a dangerous trend.

When teachers in our research studies spoke about their personal classroom development projects, they were enthusiastic and clearly committed to meeting their pupils' needs. When comparing their school's curriculum development and their own pedagogical innovations, the teachers felt that the time invested in common curricular planning should actually have been given to learners and to working in one's own classroom. The data showed that teachers had their individual ways of acting in their classrooms, which was not so much affected by written, school-based curriculum as by their personal pedagogical ideas.

The significance of a teacher's own pedagogical thinking for innovation to take place at the classroom level has been investigated by comparing the two phases of innovation: 1) the teacher-initiated, grassroots innovation wave in the 1980s, and 2) the mandated, school-based curriculum development at the beginning of the 1990s (Lauriala, 1999). A teacher who had participated in both phases, being first an active innovator in integrative teaching with two of her colleagues and then participating in the curriculum innovation as all the teachers were mandated to do, had the following to say:

> My personal opinion is that I don't believe that this kind of big wave, as curriculum work, in any way change a teacher's world or ideas, and perhaps it's never their aim . . . the small wave in the 1980s . . . has greatly influenced my whole work, and it has brought actually the kind of things that have meant a total turning point in my work. (Lauriala, 1999, p. 9)

In secondary vocational education, the stories of student teachers were very positive toward teacher autonomy, on one hand, and the interaction, positive collaboration, and sharing of ideas with colleagues on the other. They were also much more certain about curriculum and planning than primary in-service teachers, supposedly because they were still in their preservice phase and had the following to say: "I experienced the importance of planning. . . . [Careful planning] lead[s] to success in interaction with the group, which created a sense of empowerment" (Nissilä, 2005a, pp. 216–217).

Reading many teachers' narratives has convinced the authors that Finnish teachers are able to use their working practices as objects of study. It is, therefore, difficult to agree with Simola's (2005) claim concerning Finnish teachers' pedagogic conservatism. Simola actually admits having only scant empirical evidence to back up his claim. He quotes the report of the international group of assessors who evaluated a Finnish comprehensive school and who observed Finnish students working in a teacher-directed manner using a textbook, with all students doing the same thing and progressing at the same rate. Simola further found from his study that Finnish teachers mostly speak to their students as they would to adults in order to maintain order and safety. Rather than encouraging intimacy, some experienced Finnish teachers emphasize how important it is to keep a professional distance from their students, their homes, and their problems (Simola, 2005, p. 463).

Although in the 1980s there was a shift toward more learner-centered and active classrooms, Finnish schools might best be characterized as a balance of teacher control and student freedom. Students are often given autonomy to do their work, projects, and tasks that are self-chosen and self-monitored, but there are also teacher-led discussions and direct teaching. Because of the introduction of innovations based on alternative pedagogies, in the 1980s and 1990s, the strict boundaries between different subjects and class time limits had largely been broken. Thus teaching, especially in many primary classrooms, began to be integrated. Learning was understood to involve not only cognitions, but also the affective domain and social relations, now recognized as essential for effective learning (Lauriala, 1997, p. 78–79; 1995).

Unlike Sweden, for instance, all the pedagogical innovations and alternatives in Finland were carried out within the public school system. Although teachers are given pedagogical autonomy to choose their approaches and methods, learning is expected to meet the same basic standards everywhere at all levels.

The history of school development and innovation work has found that external prescriptions for practice are inefficient in leading to qualitative improvements in students' learning outcomes and experiences. Hence, it is inferred that, as professionals, teachers should be autonomous and self-regulated persons whose first and primary responsibility is to the welfare of a child or a young person. Teachers should be accountable, not to external standards, but with respect to their influence on students. On these bases, teachers' autonomous action may manifest itself through resistance to inflexible curriculum and standards set by an outsider (Lauriala, 2002, p. 128–129).

While in many other countries teachers' autonomy in school decision-making has diminished since the 1990s, the trend has been the opposite in Finland. The sectors that formerly used to dictate and order teaching, no longer assume this task, and teachers are more and more expected to define their norms of practice themselves; this includes what is good practice and what are the priorities of their school. It has not, however, been easy to change teachers' attitudes in Finland and make them act autonomously. Only changing statutes or formally declaring them autonomous is not the answer; there is a long history of cultural norms and traditions against it, both in teacher education and in schools (Lauriala, 2002, p. 109; 1997, p. 146–148; Nissilä, 1993, p. 208–210).

There are still some contradictory trends: accountability and efforts to standardize, on one hand, and striving for equality, on the other. To promote the equal access to education for the children in various parts of the country, seeking standard measures may conflict with the situation-specific solutions. Autonomy and pedagogy may be seen as interrelated and interdependent concepts, if their meanings are deeply analyzed and interpreted. To start with, there are two assumptions: 1) that there cannot be any "real," genuine pedagogy, in the proper sense of the word, without the autonomy of the teacher, without his/her freedom in decision-making and action, and 2) autonomy of a teacher does not lead to what is educationally worthwhile or educative, to what is good for students, unless it is backed by pedagogy, by a teacher's pedagogical consideration and tactfulness (Lauriala, 2002, p. 131–132; Nissilä, 2002; Korpinen, 1996).

Different teacher education departments and units have carried out several joint development projects to create novel learning environments for teachers (e.g., Kaikkonen & Kohonen, 1999; Lauriala, 1997; Nissilä, 1997; Syrjälä, Annala, & Willman, 1997). A multi-voiced impression of Finnish teachers' ways to work is apparent. Research findings highlight teachers' reflections on their ethical position, commitment, and caring (e.g. Nissilä, 2005a; Estola & Syrjälä, 2003). The concept of vocation is still applicable to many teachers' attitudes to their work (Estola, Erkkilä, & Syrjälä, 2003).

These results apply both to regular and vocational teachers, while Simola (2005) notes that his results are more indicative of teachers teaching subject matter. The subject matter teachers interviewed by Hannele Niemi (2002) were committed to using methods of active learning but were notably uncertain about their future. Niemi's overall conclusion to her extensive research was that many elements of active learning were present in Finnish schools and teacher education departments. There were, however, a number of obstacles that must be eliminated before the shift to active learning could be accomplished generally in Finland.

Discussion

The professionalization of teaching, which has been a dominant theme in educational discourse during the last two decades, is concerned with providing teachers with autonomy, privilege, and status. Characteristic of professionals is the autonomy to decide the aims and methods of their practices, not only, or even primarily, individually but collectively. Professionalization refers to shared standards of practice, a common knowledge base, and special attention to the unique needs of clients as well as responsibility for client welfare and the ability to make balanced and careful judgments (Darling-Hammond, 1994).

Contrary to the above ideal, the centralized control of teachers involves the delivery notion of professionalism. Pedagogically this is questionable and risky, because the actions of policy makers are not pedagogically or educationally based but often politically driven. In this paradigm, teachers are seen as recipients of externally determined goals and codes of action, and they are mandated to implement mechanically and uniformly the prescribed curriculum and innovations. This is accompanied by an external evaluation system aiming at the standardization of educational outcomes (Lauriala, 2002, p. 128–129). For us, the miracle of Finland is to be attributed to the pedagogical autonomy of Finnish teachers, which allows them—and us—to speak as reflective practitioners. Our high quality, research-based teacher education program gives teachers tools to be inquiry-oriented, reflective practitioners, who are oriented toward continuous development and innovation in their work.

Based on our research data, we are of the opinion that large-scale, experimental, and accountability-oriented research in teacher education and teacher practice is not enough (Richardson & Fenstermacher, 2005). Practical development requires small-scale, local, and, most often, qualitative research on teacher education and teacher development. Teachers ultimately either implement the planned changes in their own work or fail to do so. Case studies are needed, as well as self-study research, collaborative studies, and narrative inquiries to gain a deeper understanding of teachers' actions and pupils' needs, interests, and context (Lauriala, 2003).

It used to be possible for teachers in Finland to work independently. Inspectors came and offered feedback in the middle of the 1900s. Supervisors and provincial inspectors followed them. Research and development was then carried out with the support of the National Board of Education, but teachers retained the freedom to adjust and modify their teaching.

Supervision now takes the form of evaluation and national examinations, which may actually restrict teachers' freedom. These trends are contrary to the reflective practitioner principle by creating tension between the pedagogical autonomy given to teachers and schools and the external evaluation system restricting their choices. What we have found is that teachers do not want to be simply technicians who have to put into practice what the politicians and other educational authorities ask them to do (Estola, 2003; Estola & Syrjälä, 2002, pp. 184–189, 194; Lauriala, 1997). Teachers want to be involved with their whole personalities using their understanding and skills in the best interests of the learners (Nissilä, 2006). Since teachers desire to be regarded as reflective practitioners and professionals, they should be given autonomy to do their work.

Our own studies represent qualitative case studies, narrative inquiries, action research, and interpretive interaction research. However, we recognize that teacher education in Finland needs fine-grained research and comprehensive quantitative studies equally. That way we will describe, explain, understand, and correct things over time and within and across contexts. The researcher collaborating with teachers is presupposed to conceptualizing the phenomena and experiences that teachers describe, and to analyze critically the present situation while articulating the call for transformative education (Lauriala, 2003).

Recommendations and Challenges

In Finland we are immersed in the sociological orientation of the Nordic welfare system. According to this system, welfare relates to individual resource management. The follow-up to the welfare development has been a resource-oriented concept of welfare, although we underline the fact that state provisions consist of more resources than the material side alone (Heikkilä & Kautto, 2004, p. 12). It seems that the concept of resource-based welfare together with individual experiences and orientations will be successful. Information of this kind can open up new windows to teacher education. Not only statistics but personal experiences are important when we approach issues that easily go ignored, such as gender, ethnicity, and language (Gordon, Holland, & Lahelma 2000; Sunnari, Kangasvuo, & Heikkinen, 2002; Nissilä, 2005b). Despite the miracle of PISA, not every student is happy in Finland. A major challenge is to develop all Finnish schools into places where every student can feel accepted.

Although there is a great concern for education in Finland, schools are now provided with less funding than in the early 1990s. Small schools are disappearing, which can lead to a displacement of teachers and a decrease of teachers' motivation and commitment. We do hope that decision makers and other community leaders will recognize the critical role of education in the future of institutions, communities, and nations, and fight for it. In providing for the future, teacher education is situated in a key position. Our studies show evidence that the quality of our school of education depends on our ability to take seriously the scholarship of practice and to cultivate a dialectical relationship between research and practice (Lauriala, 2003; 1997, p. 152; Nissilä, 2006).

The interconnections and tensions between the social, economic, and political structures as well as the day-to-day life and concerns in Finland are important to chart and to make visible in order to aid teachers and politicians in understanding their influences. Teacher autonomy and local decision-making in school affairs are emphasized in Finland presently. We should also consider how the system demands and changes the framework as well as constrains teacher practices. For instance, assessment practices—which seem to be the focus of teachers and schools, as well as university development at the moment in Finland—are subject to changing system demands related to tendencies toward increasing monetarization, marketization, and bureaucratization within the schooling system (Lauriala, 1997, p. 152). The increased testing of competencies is determined by the specific interests of economic enterprises. Such tendencies are not objectionable in themselves, but they may become objectionable if it were to be shown that they are educationally dysfunctional (Kemmis, 1994). However, it seems that, in Finland, teachers can still concentrate more on pedagogical issues and work for the best of the learners. Their work is not as bounded by external, standardized testing measures.

From our research and our knowledge of the challenges and problems confronting education today, we argue for action-oriented kinds of research that would involve not only teachers, student teachers, and researchers, but student learners in schools as well. To fight *against* the marginalization and *for* the vitality of local communities, multi-disciplinary approaches that call for the participation of parents, local decision makers, teachers, and researchers are recommended. The bulk of research funding should be directed toward these needs.

Authors' note: We would like to thank Professor Philip Gammage for his helpful suggestions concerning the comparative issues presented in this chapter.

References

Beaton, A. E., & O'Dwyer, L. M. (2002). Separating school, classroom and student variances and their relationship to socio-economic status. In D. F. Robitaille & A. E. Beaton (Eds.) *Secondary analysis of the TIMSS data* (pp. 211–231). Dordrecht: Kluwer.

Darling-Hammond, L. (Ed.) (1994). *Professional development of schools: Schools for a developing profession.* New York: Teachers' College Press.

Estola, E. (2003). Hope as work—Student teachers constructing their narrative identities. *Scandinavian Journal of Educational Research, 47*(2), 181–203.
Estola, E., Erkkilä, R., & Syrjälä, L. (2003). A moral voice of vocation in teachers' narratives. *Teachers and Teaching: Theory and Practice*, 239–256.
Estola, E., & Syrjälä, L. (2002). Whose reform? Teacher's voices from silence. In R. Huttunen, H. Heikkinen, & L. Syrjälä (Eds.), *Narrative research* (pp. 177–195). *Voices of teachers and philosophers*. Jyväskylä: Sophi.
Goodson, I. V., & Numan, U. (2002). Teacher's life worlds, agency and policy contexts. *Teachers and Teaching: Theory and Practice, 8*(3), 269–277.
Gordon, T., Holland, J., & Lahelma, E. (2000). *Making spaces: Citizenship and difference in schools.* Houndmills: Macmillan Press Ltd.
Hansen, D. T. (1998). The moral is in the practice. *Teaching and Teacher Education, 14*(6), 643–655.
Heikkilä, M., & Kautto, M. (Eds.) (2004). *Welfare in Finland.* STAKES. Helsinki: National Research and Development Centre for Welfare and Health.
Kaikkonen, P., & Kohonen, V. (Eds.) (1999). *Elävä opetussuunnitelma. Osa 2.* Opettajien ääniä: opettajat tutkivat työtään. Tampereen yliopiston opettajankoulutuslaitoksen julkaisuja 18. (Living Curriculum. Part 2. Teacher voices: teachers study their work. Tampere: Publications of the Department of Teacher Education, no. 18.)
Kemmis, Z. (1994). *School reform in the '90s: Reclaiming social justice.* Paper presented at the conference "Touchstones of the socially-just school," Flinders' Institute for the Study of Teaching, South Australia.
Kess, H. (2002). The child's right to early childhood services in Finland. In K. S. Chan Loran & E. J. Mellor (Eds.), *International Developments in Early Childhood Services. Rethinking Childhood, vol. 26* (pp. 71–79). New York: Peter Lang.
Korpinen, E. (1996). *Opettajuutta etsimässä.* Kunnallisalan kehittämissäätiö. Helsinki: Polemia-sarjan julkaisuja 18. (Searching for Teachership. Publications of The Development Fund of Municipalities, no. 18.)
Kupari, P.,Välijärvi, J., Linnankylä, P., Reinikainen, P., Brunell, V., Leino, K., & Sulkunen, S. (2004). *Nuoret osaajat. PISA 2003—tutkimuksen ensituloksia.* Koulutuksen tutkimuslaitos. Jyväskylä: Korpijyvä. (Young Achievers. Preliminary Results of PISA 2003 Research. The Research Institute of Education. Jyväskylä: Korpijyvä.)
Lauriala, A. (2003). *Changes in research paradigms and their impact on teachers and teaching.* Paper presented at the 10th Anniversary Conference of the Pedagogical University of Tallinn, Estonia.
Lauriala, A. (2002). Teacher autonomy and pedagogy. In T. Kuurme & S. Priimägi (Eds.) *Competing for the future: Education in contemporary societies* (pp. 127–146). Contributions of the Colloquium of the European Forum for Freedom in Education. Tallinn: Tallinn Pedagogical University.
Lauriala, A. (1999). *Professional development, school improvement and teacher autonomy: Teachers' perspectives on autonomy in two different school political phases in Finland.* Paper presented at the 9th ISATT Conference on 27–30 July 1999, Dublin, Ireland, 13.
Lauriala, A. (1997). *Development and change of professional cognitions and action orientations of Finnish teachers.* Dissertation. Acta Universitatis Ouluensis. E Scientiae rerum socialium 1997, 27, 315 s.

Lauriala, A. (1995). Student teaching in a different environment. Examining the development of students' craft knowledge in the framework of interactionist approach to teacher socialization. University of Oulu, Reasearch Reports of Department of Education no. 95.

Lauriala, A. (1992). The impact of innovative pedagogy on teacher thinking and action: A case study on an in-service course for teachers in integrated teaching. *Teaching and Teacher Education, 5*(6), 523–536.

Lauriala, A., & Syrjälä, L. (1995). Influences of research into alternative pedagogies on the professional development of prospective teachers. *Teachers and Teaching: Theory and Practice 1*, 101–117.

Lindblad, S., Johannesson, I., & Simola, H. (2002). Education governance in transition: an introduction. *Scandinavian Journal of Educational Research, 46*(3), 237–245.

Linnakylä, Pirjo (2004). Finnish education—Reaching high quality and promoting equity. *Education Review, 17*(2), 35–41.

Moore, E., Antikainen, A., & Kosonen, T. (2005). Research review of restucturing in health and education in Finland. In I. Goodson & C. Norrie (Eds.), *A Literature Review of Welfare State Restructuring in Education and Health Care in European Contexts: Implications for the Teaching and Nursing and their Professional Knowledge* (pp. 25–47). EU Sixth Framework Programme.

Niemi, H. (2002). School experiences and moral orientation among Finnish and Chinese adolescence. A paper presented at the Education and Cultural Diversities NERA's 30th Congress, 7–9 March 2002. Tallinn, Estonia.

Niemi, H. (1988). *Is teaching also a moral craft for secondary school teachers? Cognitive and emotional processes of student teachers in professional development during teacher education.* University of Helsinki. Department of Teacher Education. Research Report 61.

Nissilä, S-P. (2006). *Dynamic dialogue in educational contexts. Towards transformation in vocational teacher education.* Unpublished doctoral dissertation submitted for publication. Tampere: Tampere University.

Nissilä, S-P. (2005a). Individual and collective reflection: How to meet the needs of development in teaching. *European Journal of Teacher Education, 28*(2), 209–219.

Nissilä, S-P. (2005b). *Teacher students' prior learning experiences.* A paper presented at the Congress of International Study Association of Teachers and Teaching on 4/7/2005 in Sydney, Australia.

Nissilä, S-P. (2002). Pratique reflexive et besoin dáutonomie dans la formation des enseignants. A. Camilleri (Ed.), *Introduction de l'autonomie de l'apprenant dans la formation des enseignants* (pp. 11–19). Strassbourg: Editions du Conseil de l'Europe.

Nissilä, S-P. (1997). Raising cultural awareness among foreign language teacher trainees. In M. Byram & G. Zarate (Eds.) *The sociocultural and intercultural dimension of language learning and teaching* (pp. 55–72). Srassbourg: Council of Europe Publishing.

Nissilä, S-P. (1993). *Kielenopettajan oppimiskäsityksen tausta* (The Origins of the Learning Conceptions of Foreign Language Teachers). Licentiate research. Oulu: The Faculty of Education.

OECD (2004a). *Learning for tomorrow's world. First results from PISA 2003.* Paris: OECD.

OECD (2004b). *Education at a glance. OECD indicators 2004.* Paris: OECD.

OECD (2001). *Knowledge and skills for life. First results from PISA 2000.* Paris: OECD.

OECD (2000). *Early Childhood Education and Care Policy in Finland.* Background report prepared for the OECD Thematic Review of Early Childhood Education and Care Policy. www.oecd.org/document/49/0,2340,en_2649_34511_1941745_1_1_1_1,00.html. Retrieved January 26, 2006.

Richardson, V., & Fenstermacher, G. (2005). Research and the improvement of practice and policy in teacher education. *Didacta Variea, 10*(2), 7–26.

Räsänen, R. (2002). Toimintatutkimuksen kautta interkulttuuriseen opettajankoulutukseen Oulun yliopistossa (Through action research to intercultural teacher education in the University of Oulu). In R. Rasanen, K. Jokikokko, M-L. Järvelä, & T. Lamminmäki-Kärkkäinen (Eds.) *Interkulttuurinen opettajankoulutus. Utopiasta todellisuudeksi toimintatutkimuksen avulla* (Intercultural Teacher Education. From Utopia to Reality with the Help of Action Research) (pp. 117–137). Acta Universitatis Ouluensis E55.

Siljander, P. (2006). *How to explain the school achievement? A historical analysis of factors behind PISA-success of Finland.* Paper presented at the 4th Annual Hawaii International Conference on Education, 6–9 January 2006, Honolulu, Hawaii.

Simola, H. (2005). The Finnish miracle of PISA: Historical and sociological remarks on teaching and teacher education. *Comparative Education, 41*(4), 455–470.

Sunnari, V., Kangasvuo, J., & Heikkinen, M. (Eds.) (2002). Gendered and sexualised violence in educational environments. *Femina Borealis 6.* Oulu University Press.

Syrjälä, L. (1998). Action research, teacher education, and school practice. In R. Erkkilä, A. Willman, & L. Syrjälä (Eds.) *Promoting teachers' personal and professional growth* (pp. 9–22). Acta Universitatis Ouluensis E32.

Syrjälä, L., Annala, H., & Willman, A. (1997). *Arviointi ja yhteistyö koulun kehittämisessä. Oulun opettajankoulutuslaitoksen ja Oulun kaupungin koulujen evaluaatioprojektin lähtökohtien ja alkuvaiheiden kuvausta.* Oulun yliopiston kasvatustieteiden tiedekunnan opetusmonisteita ja selosteita 73 (Evaluation and cooperation in developing schools. Description of the starting points and phases of the evaluation project in the schools of the city of Oulu. The reports of the Educational Department of Oulu University 73).

Välimäki, A-L. (1998). *Päivittäin. Lasten Päivähoitojärjestelyjen muotoutuminen varhaiskasvatuksen ympäristönä suomalaisessa yhteiskunnassa 1800- ja 1900-luvulla* (Every day. The evolution of the children's day-care system as an environment for growth in Finnish society in the 19th and 20th centuries). *Scientia Rerum Socialium,* Acta Universitatis Ouluensis, E31.

Westbury, I., Hansen, S-E., Kansanen P., & Björkvist, O. (2005). Teacher education for research-based practice in expanded roles: Finland's experience. *Scandinavian Journal of Educational Research, 49*(5), 475–485.

Yang, Y. (2003). Dimensions of socio-economic status and their relationship to mathematics and science achievement at individual and collective level. *Scandinavian Journal of Educational Research, 47*(1), 21–41.

CHAPTER 13

Creating a Learning Space for Educators
POLICY DEVELOPMENT FOR ACCOUNTABILITY SYSTEMS

Ora W. Y. Kwo
The University of Hong Kong

> Ora W. Y. Kwo, Ph.D., is an associate professor of education at the University of Hong Kong. As a university academic involved in teacher education for over 20 years, she specializes in research on processes of learning to teach and professional development. In 1997, she was awarded a University Teaching Fellowship by the University of Hong Kong in formal recognition of her excellence in teaching. Since then, her research interests have extended to the quality of teaching and learning in higher education, and the building of learning communities.

ABSTRACT

> This paper presents an interpretation of the impact of accountability systems on teachers' professional development from a Hong Kong perspective. It begins with factors that influenced the accountability policy on teachers since Hong Kong's change of sovereignty in 1997. By interpreting a government-funded study in the light of relevant literature, questions are raised about co-ordination between policy tracks, alignment of expectations from policy and professional workers, and the meaning of professional autonomy. Interviews on school educators' professional responses to accountability expectations are then reported. Paradoxically, with the intention to bring about enhanced performance rooted in autonomous learning, accountability systems are often taken by policy-makers as an end that results in teachers feeling consumed by paper work. The chapter concludes with an assertion of the need to create a learning space and re-establish relationships among educational stakeholders in a

professional environment where teachers can be truly liberated from accountability pressures.

Rarely are societies content with their education systems, particularly under accelerated processes of social change that are shaped not only by local forces but also by wider forces of globalization. Indeed, it seems that almost everywhere both macro- and micro-level administrators are required to be constantly alert to ways in which their systems and institutions can be shaped to improve competitiveness, responsiveness, and leadership in evolving social and economic circumstances. Hong Kong has certainly been at the forefront of international developments, especially in finance and trade.

Society in Hong Kong must also take account of domestic and regional changes. Domestic changes particularly arise from the reversion of sovereignty over Hong Kong from the United Kingdom to China in 1997. The re-integration with China set out a new era for the post-1997 government. The first Chief Executive, Mr Tung Chee-Hwa (1997), declared that education was one of his three major concerns. Reform proposals were set up (Education Commission, 2000), from which stemmed the School Development and Accountability (SDA) policy in 2003.

Accountability measures were accompanied by a generic teacher competencies framework aiming at the improvement of the professional quality of teachers (ACTEQ, 2003). These reform initiatives, which followed changes in government, had many parallels elsewhere (Mebrahtu et al., 2000; Bray & Lee, 2001). In many respects, therefore, the specific context on which this paper is grounded also had parallels elsewhere and had features that would be widely recognizable. This chapter presents an interpretation of the impact of accountability systems on teachers' professional development from a Hong Kong perspective. Built on documentary analysis and empirical data from case studies, it reveals tensions between accountability and autonomy, and asserts the need for the creation of learning space among educators and stakeholders.

Background to Accountability Policy on Hong Kong Teachers Since 1997

Following the 1997 political transition, the Education Commission, which was first set up in 1984 as a non-statutory body to advise the Government on the overall development of education in light of community's needs took on the responsibility for coordinating various major executive and advisory initiatives for educational reform. Issues relating to the interface between different education sectors were dealt with through committees and publications, and the progress has been reported to the public regularly through a website (www.e-c.edu.hk). It was summarized in the 1st Progress Report (Education Commission, 2002, Chapter 1) as follows:

> To meet the needs of the society in the 21st century, the Education Commission in 1998 embarked on a two-year comprehensive review of the overall education system in Hong Kong. The review covered curricula, academic structure and assessment mechanisms at various stages of education, as well as the interface between the different stages. Following three rounds of extensive public consultation, a series of recommendations for refining the education system was submitted to the government in 2000. The Chief Executive endorsed all the recommendations and announced a timetable for implementing the reform measures. (p. 3)

Upon further analysis of the three Progress Reports on Education Reform (Education Commission, 2002, 2003, 2004), the following major claims of progress were identified:

- Learning has become more interesting for students because of the diversified curricula.
- School principals and teachers have been given greater autonomy in designing curriculum and teaching strategies.
- The school system has become more diversified for the community at large.
- Transparency of the reform process has been enhanced to allow the public to have a better understanding of the timetable and measure of the reform.

The third Progress Report (Education Commission, 2004, chapter 3), however, also suggests that there are challenges in implementation. The nature of challenges can be inferred from the statement of "the way forward" as a continual quest for:

- comprehensive and systematic research to understand the impact of various reform measures
- support and cooperation of different stakeholders
- statistics to allow assessment of the effectiveness of the reform
- professional-driven culture of self-improvement
- change from a highly centralized system to a school-based model of support

In addition to the advisory Education Commission, the government's Education and Manpower Bureau is responsible for the formulation of policies, introduction of legislation, and implementation of reform programs as described in its website (www.emb.gov.hk). In the reform programs was a move toward formalized measures in accountability. The School Development and Accountability (SDA) policy was first introduced in the 2003/04 school year under the assumptions that (a) school self-evaluation is an internal quality assurance mechanism to enhance school development and accountability, and (b) external school review enhances school development and accountability. By July 2005, the External School Reviews (ESRs) had been conducted in 249 schools, with School Self-Evaluation (SSE) taken as part of the function to serve school development and improvement.

Impact from Implementation of Accountability Systems

Since the policy implementation, research-based knowledge of the impact of the External School Review (ESR) and School Self-Evaluation (SSE) as accountability systems has been made available on the website www.emb.gov.hk that can be traced from kindergarten, primary, and secondary education under the section "Initiatives Highlights." The initial impact can primarily be interpreted from a report by an independent research team from Cambridge, United Kingdom (MacBeath & Clark, 2005). Funded by the Education and Manpower Bureau to evaluate the Phase I Implementation of SSE and ESR in the 99 schools involved, the study drew on comprehensive quantitative data from eight case studies and eleven focus group interviews. According to the report, the impact was identified in five key areas:

- leadership and management
- school culture
- teaching and learning
- professional development
- self-evaluation

Among the schools being studied, it was revealed that the central concern was the apparent compliance to the running of the systems. On the other hand, little was known about the in-depth responses to what the systems were intended to do in terms of the beneficiaries. While leadership and management gave testimony to changes seen in a more distributed or inclusive approach, and while a more reflective school culture has been induced with attention to student-centered pedagogy and teacher professional development, it is difficult to see beyond a performance-orientation approach. As summarized in the report: "What emerges very clearly is that impact exists in myriad ways but the extent to which schools own internal capacity for self-evaluation has been enhanced remains a more open question" (MacBeath & Clark, 2005, Paragraph 19.5).

In conclusion (Paragraph 20), the study recognized clear evidence of objectives being achieved in terms of:

- a deepening understanding of the purposes of ESR and SSE
- promoting the use of data and evidence as a basis for SSE
- conducting informed discussions as to the value of the SSE and its relationship to school improvement
- getting better at identifying their strengths and areas for improvement
- developing a more systematic and informed approach to SSE
- creating a greater sense of openness and transparency
- incorporating different stakeholders' perspectives on the relationship between ESR and SSE

It may be natural that within a relatively short period since the introduction of formal accountability systems, schools are taking time to consider how they present themselves regardless of whether they are able to authentically maintain the congruence between the private aspects of reflectivity and the public aspects of desirable performances. As teachers are expected to be the frontline agents for educational reforms, it is essential to understand the truthful inner picture about their capacity and sustainability in adapting to changes that impact on their *learning and development*. This can be detected in the report, with themes in italics outlined by the author:

Sustainability. Most of the case study schools had experience of working with higher education teams, projects which had raised awareness and identified issues, but generally failing to provide evidence of a sustained momentum. (Paragraph 2)

Teachers' perception of the purpose for SSE. SSE brought a new sense of urgency to the development of self-evaluation. It tended not to be seen, however, as an extension or refinement of what went before but rather as another new initiative. At this stage of development there are still many teachers who see the primary audience for SSE as the review team rather than for the school itself. (Paragraph 3)

There remains a question for many school staffs as to who self-evaluation and review are for and who should have access to the review team report. In some cases teachers remained uninformed while in others teachers had been informed and included in all stages. (Paragraph 14)

Teachers' perception of workload. At present, there remains a widespread view that SSE adds to workload. This is an indicator of the extent to which SSE is perceived as an extra, imposed rather than owned. (Paragraph 4)

Orientation to self-evaluation. Some schools already had an embryonic self-evaluation culture while others had found external review a catalyst for developing it. (Paragraph 6)

Perspectives of stakeholders. The press for evidence has helped schools move from a more subjective and impressionistic evaluation of their own performance to a more systematic and rigorous approach to assessing the quality of practice. The inclusion of a range of stakeholders in the process has encouraged schools to view their practice through different lenses and has challenged complacency and self-satisfaction where that existed. (Paragraph 9)

Teachers' confidence and support needed. Overall, responses suggest that teachers are happiest when it comes to issues of information and understanding and least happy when it impacts directly on their work or professional lives. The most negative responses were in relation to the personal, emotional, and professional impact of ESR . . . Lack of confidence in the use of self-evaluation tools and commitment to learning more in this domain also highlight priorities for further development. (Paragraph 15)

The process of discussing the 14 areas in the SSA contributed significantly to this greater openness and sharing. While for many staff it demonstrated that 360 degree evaluation need not be a threat for others there will be a need for continuing support in dealing with critical feedback. (Paragraph 17)

Relationship between SSE and ESR. This study identified factors that stand in the way and those that promote an effective relationship between SSE and ESR. Those that were most consistently cited as promoting that relationship were: building confidence, giving impetus to cultures of self-evaluation, enhancing school development, promoting a positive view of ESR . . . Factors cited as inhibiting were questions of purpose, apprehension, and vulnerability, time, the expertise of the review team. (Paragraph 18)

The relationships between the observations can be conceptualized as in figure 13.1. The central concern of the impact of new initiatives can be identified as teachers' sustainability in adaptation to change. This is inevitably affected by teachers' perception of the intended purpose and the implications on workload. Whether teachers perceive the accountability systems as creating extra demands or as integral parts of daily professional activities depends on the culture of the environment in which mutually-related teacher factors include: orientation to self-evaluation, objectivity to consider stakeholders' perspectives, confidence and support needed, and established relationships between SSE and ESR.

In essence, the newly emerged accountability systems of SSE and ESR were implemented at a brisk pace in Hong Kong. This initial impact study has added

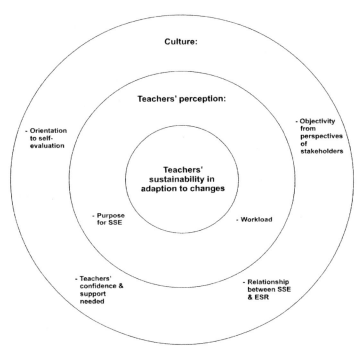

Figure 13.1 Conceptualization of teachers' sustainability in adaptation to changes

to the transparency of the processes and outcomes, and facilitated understanding of the challenges and dynamics of educational reform initiatives. However, the challenges remain, with the "tendency among schools to relax following the review, treating ESR as an end point rather than a beginning" (MacBeath & Clark, 2005, p. 9). Indeed, some schools may not have been doing what the Education and Manpower Bureau required. According to a letter dated July 29, 2005, issued by the Quality Assurance Division to schools:

> There have been complaints about excessive workload and documentation in preparing for the ESR. For instance, we have received a school self-assessment (SSA) report containing more than 100 pages and some ESR teams have been presented with as many as 72 items of recently prepared documentation, including detailed minutes of the array of meetings in schools. Some schools spent an inordinate amount of time rehearsing and coaching for the ESR. This is totally unnecessary.

As reflected in the above statement, the "misunderstanding" from some schools was not just related to procedural matters. The excessive effort in presenting performance details may have been grounded in a perception that the review was a means of checking whether the schools were "good enough" when faced with survival threats, given the decline of the student population in Hong Kong. This contextual factor was not addressed in the impact study. The appropriateness and sustainability of efforts in responsive commitment to accountability systems could only be addressed beyond the realm of this reported impact study by confronting the core meaning of the reform to all stakeholders.

Policy Formation and Development

Reform initiatives with specific timetables tend to assume a linear pattern of well-shaped policy decisions followed by programs leading to changes in practice for specific outcomes. In the Hong Kong scenario, the government has adopted a firm commitment to tight timetables for reform (see Education Commission, 1997, pp. 50–58; Education and Manpower Bureau, 2005, p. 144). The understanding of the change process is critical, but cannot be left merely to rounds of public consultation. The reality in reform processes goes beyond means-ends approaches to rationality, a sense of chaos resulting from the unknown seems to prevail. As pointed out by Levin (2001):

> Human abilities to understand problems and generate appropriate solutions are limited and often inadequate to the complexity of the problems. The entire process of policy development and implementation takes place in a context that is constantly changing, multifaceted, and very difficult to read. . . . [Strategies for reform tend to focus on] what can be done instead of what might really make a difference. . . . History and culture are very powerful influences on policy and practice. (p. 23)

As pointed out earlier, challenges in the implementation of education reform in Hong Kong were recognized in the 2004 Progress Report (Education Commission, 2004, Chapter 3). Embedded in the statement for the way forward is a gap between intended and actual outcomes. This calls for collaboration among the different stakeholders in the reform agendas. Indeed, the nature of this "collaboration" from different sectors is not yet understood and recognized as the central focus for concerted effort, without which the sustainability of pacing will be in question. Meanwhile, as a further interpretation of the reform journey ahead, policy formation and development can be queried in relation to school educators' responses.

Coordination between policy tracks. With a steady flow of reform documents published by the Government since 1997 (see websites of Education Commission and the Education and Manpower Bureau), the intriguing question concerns the different paces inherent in the timelines of the government and the practitioners on whom the success of reform initiatives depends. While each of the reform policy tracks may sound convincing, and demands professional response within an externally imposed time-frame, altogether the reform agendas may present a complex scenario of huge demands for which school educators need to re-construct their sense of orientation. The questions arise: Is coordination between policy tracks part of the agenda at the level of policy-formation? At the level of implementation, how do school educators respond professionally? How do they respectively construct strategies in coherence so that the intended beneficiaries (i.e. the students) will gain accordingly? More specifically, how do school principals and teachers work together in the contexts of respective school traditions toward the intended outcomes?

Alignment of expectations from policy and professional workers. The process from policy making to policy implementation may seem logical. However, governments in open political systems have limited ability to create the world as intended. According to Levin (2001), reforms are influenced by groups outside of government, and policy ideas come from the civil service and from those engaged in education, all of whom influence practical politics and institutional structures. Official and unofficial checks and balances affect not only what policies are adopted, but even more, what happens to policies in the move from adoption to implementation (p. 191). If Hong Kong is to maintain its openness in policy development, the influences outside the government are part of the processes for changes. Whatever its limits, public policy remains a central way for societies to shape themselves. Alignment of expectations from policy and professional workers becomes a major challenge for policy development.

Meaning of professional autonomy. When school educators are expected to be the primary agents for educational reforms, a fundamental question concerns whether professional autonomy is respected and enhanced in the dynamics of the reform movement. A preliminary consideration is about the initiation and ownership of educational reform. In Hong Kong, as the initiator of official reform agendas, the Education Commission is a body with membership of community elites, despite its limited representation as a decision-making establishment in initiating changes for diversified frontline practitioners. The meaning of professional autonomy is particularly critical in various school contexts. How do frontline professionals share the ownership of accountability systems in terms of concept and professional practice? Given that the reform initiation from the Education Commission is a foundation for public attention, the ownership of reforms from educators and stakeholders should be a continuing process of scaffolding to be continually built.

To move beyond the impact study by MacBeath and Clark (2005) that focuses on the effectiveness of the accountability systems, this chapter raises concerns about *how* the accountability systems affect school educators' autonomy and explores how the accountability systems are being taken professionally as a means toward the educational goals that they are meant to serve.

Re-focusing on School Educators

With a view that accountability systems are set up to measure and recognize achievements that are grounded in professional autonomy in everyday practice, it is through the high-scoring schools in accountability records that such autonomy can be best understood. In this light, school educators' voices about professional autonomy were solicited from two schools, which were selected for their positive external reviews and their various lengths of history. One was recently set up after the turn of the century, and the other had a history of almost four decades. In each setting, interviews were conducted with the principal and a sample of four teachers with a range of seniority and from different curriculum areas. The interviews focused on their conceptual understanding of and professional responses to accountability demands in terms of the following questions:

- What is your understanding of the development in professional autonomy in the context of the accountability systems you have experienced?
- As a school with a positive external review, what do you see as major strategies attributing to the development of autonomy? What problems can you identify, and how do you overcome them?

The major goal for this study was to shed light on documentary analysis of government reform initiatives and the impact study, as set out in the earlier parts of this chapter. Voices from school educators were solicited to identify critical dimensions for policy development and educators' choices. Data from the interviews were collated for analysis and derivation of common perceptions among school principals and teachers.

PRINCIPALS' VOICES

Both principals were positive about the rationale for the educational reforms. They appreciated the significance of accountability systems and were ready to incorporate them into their schools. They especially believed in the significance of continuous professional development for their teaching staff to cope with educational changes; yet, at the same time, they were aware of the time pressures on teachers. Their views on accountability were rather similar, and they both believed that teachers had to be accountable to themselves rather than to external forces.

There was a shared concern about the emphasis on measurable outcomes, given that many time-consuming efforts without immediate outcomes cannot easily be measured. While being open to questioning the validity of the indicators for measurement, one school principal had actually proceeded to apply for funds to design an alternative performance framework for optimizing teachers' professional standards. According to the two principals, teachers' practice in meeting accountability demands must be taken into consideration along with the claim of professional autonomy:

"Everything must start from within each person. Accountability systems are important, but at the same time, teachers must engage in professional practice with a sense of autonomy. The two must exist in harmony."

"If teachers need to be monitored, the reform initiatives are not working."

"Educators must be entrusted with autonomy to make choices based on professional knowledge, with awareness of the consequences of the choices and willingness to take the consequences."

Both principals recognized the limitations of initial teacher education and one-shot external training programs. They saw the value of initial teacher education as much more than the curriculum content, as they wanted novice teachers to be prepared with a vision of long-term goals in education. Respectively, both principals provided spaces for their own teaching staffs to run school-based programs with constant support and praise, instead of relying on staff development training from outside "experts." They considered collegial input as valuable resources and held high regard for relevance of 'training' and continuity of effort on shared problems and the building of contextual knowledge when addressing them.

Aspiring for a learning culture, these school leaders constantly encouraged critical discussion in the supportive school environments, with staff development being prioritized over staff assessment. In addition, they were ready to promote opportunities for faculty to engage in professional interactions with other schools and occasionally serve as guest speakers. Their articulation of further developmental strategies included cross-curricular lesson observation, mentoring support for induction of new teachers, and systematic involvement of students in providing feedback to teachers. Both principals had strong confidence in the capability of their teaching staff and students and were inclined to think holistically in building the learning environment.

TEACHERS' VOICES

In both schools, the teachers were generally aware of changes in society and around the world and were ready to engage in education reform. However, they felt too much was expected within a short time and that they simply did not have enough time to live up to the multiple expectations. However, most teachers claimed that the accountability systems had placed little pressure on them. They perceived that, as professionals, they were responsible to themselves and their students, whether or not there were any external school review exercises. It also was important to them that the principals did not pressure them. Some teachers reported that accountability systems may serve as a means to monitor less conscientious teachers. The real pressure appeared to come from keeping detailed records on various domains of their commitments, as they perceived that the amount of time required could be more desirably spent on students.

In considering the Generic Teacher Competencies Framework, most teachers accepted it as a reminder to reflect in various domains. Of the three levels (Threshold, Competent, and Accomplished), most teachers felt they had reached at least the competent level. Many identified themselves at the accomplished level in the Teaching and Learning Domain, but considered themselves weakest in the Professional Relationships and Services Domain.

With respect to engagement in continuing professional development, most recognized the need, not because of pressure from accountability systems but due to awareness of the changing society in which the updating of knowledge and skills was considered essential. Accordingly, many selected courses that matched their needs. Paradoxically, some still claimed to attend courses for the sole purpose of satisfying the stipulated requirement of the Education and Manpower Bureau for engaging in 150 hours of continuous professional development over a three-year period. They regretted having to rob time from other professional commitments, let alone their families. Their suggestions for further staff development included: school-based staff development programs for small groups with specialized topics and further exchanges with other schools for wider exposure to various contexts for professional development.

HARMONY OF VOICE

The interview data indicated that school educators with a clear sense of autonomy were generally receptive to the need for adapting to changes of the world and appreciated the value of continuing professional development. The school principals and teachers in both schools were in favor of school-based continuing professional development activities through which they could actualize professional development based on autonomous follow-up practices. At the same time, they found that the external courses provided by the Education and Manpower Bureau were not as practically relevant to meet their needs. In essence, for any school to promote a staff development culture, the leadership role of the principal was most significant. Such leadership could be defined in terms of a broad and long-term vision from which to motivate the teaching staff to claim autonomy while offering understanding support.

With numerous reform initiatives being introduced simultaneously, time constraint was a well-cited factor that posed a threat to the harmony between accountability and autonomy. Paradoxically, even when accountability was intended to bring about the impact of enhanced performance rooted in autonomous learning and development, the system could be taken as an end itself by policy-makers and leave teachers feeling consumed by paper work. Yet, it was the learning autonomy of dynamic school educators that affirmed professional sense of prioritizing in time management. School leadership maintained a "buffer zone" between external pressures and the teachers, as comprehension of the rationale behind each initiative enabled decisions relating to priority.

An understanding of the strengths and weaknesses of the teaching staff was vital. Teamwork had to be promoted in an atmosphere of openness and trust. In essence, a supportive environment was critically important for nurturing the learner-teachers. This translated into integrating the ongoing learning into teachers' daily professional lives as a "habit" rather than a separate chore arising from the mandates of the accountability systems.

Paradoxes on Perceptions of Relationships

Amidst reform initiatives, accountability systems are intended to hold educators responsible for improvement and better service. This chapter has explored the meaning of autonomy from the perspectives and lived experiences of school educators who have a clear sense of autonomy. Moreover, it reached an understanding that an improved educational service must be grounded in the recognition of the nature of the school environment and educators' ownership of reform. However, this sense of ownership is not given. It has to be claimed with persistence and autonomy, given the official perception of the relationship between stakeholders and the government. According to the Education Commission (2002), the government clearly takes up the roles in initiation and implementation.

In addition to coming from the government, expectations of teachers and students come from a myriad of stakeholders including cultural organizations, the commercial sector, parents, youth services groups, teacher training providers and curriculum experts, and school sponsoring bodies. According to this perception, education reform is to be "implemented" with "concerted effort" in terms of "communication and partnerships." Likewise, the concepts of "support" and "cooperation" are centered upon the agenda presented by the government, and "autonomy" is known as being "provided to" school principals and teachers in designing curriculum and teaching strategies. While ensuring accountability of school educators to *carry out* the reform, the government becomes the bearer of accountability of the reform itself, as claimed:

> For transparency of the reform implementation, we will report progress to the public and exchange views with various stakeholders on the reform implementation on a regular basis. (Education Commission, 2002, p. 36)

Accountability, then, becomes a reporting system for the sake of transparency, and the focus is placed on allowing the public to have a better understanding of the timetable and measure of the reform. With an emphasis on transparency, the government can, at best, design a model of proving to the public its abundant efforts in urging reform while expecting school educators to be responsive through accounting activities in terms of reform implementation that is externally defined.

Accordingly, the leadership role for accountability is strongly associated with the promotion of a performance-oriented culture. Yet, the professionalism that the accountability systems are intended to induce and assure goes beyond the visibly quantifiable accounts. Could the ways that the systems are perceived or run in the majority of schools encourage excessive focus on the threshold? Could the current official perception of the government's ownership of the reform agenda lead to fragmentation of responses for the sake of meeting the transparency demands? Could the emphasis on public transparency paradoxically steer attention away from the destiny for the public good?

Alternatively, could a momentum of pursuing higher goals beyond the threshold of the performance-oriented frame be sustained and recognized? Above all, with the intensification of reform agendas externally initiated, can school educators guard against the dissipation of professional energies away from students—those who are meant to be the beneficiaries? These questions cannot be addressed in isolation. To take a holistic stance, all stakeholders must converge and focus on creation and development of a learning space.

Toward a Renewed Relationship with Convergent Focus on Learning

This chapter is now ready to argue for a re-orientation of relationships between the government and the stakeholders, with a convergent focus on learning. A learning space needs to be created for this re-orientation, from which stakeholders can coordinate their efforts for joint accountability to the public.

Autonomy is to be claimed by educational professionals and respected by policy-makers—not to be replaced by the systems, however justifiable the systems are. In parallel, change and refinement of policy and systems are desirable traits of learning for which the government officials must be given the necessary space. In the context of intensification, learning space and time must be owned and protected by the professionals as a major form of autonomy. Support is a form of mutuality rather than a one-way relationship that may be experienced as an imposition, particularly when the ownership of the meaning of the function is in question.

Adapted from Palmer (1998, p. 102) and concluded in a research and development project on teacher induction and continuing professional development that was funded by the Education and Manpower Bureau (Kwo, 2007), this chapter asserts a collective quest for the community of truth for school educators. A learning space can be created from among different schools as different communities, far-flung across space and ever changing through time. At the center of this learning space, there is always a subject. This relationship begins when the subjects occupy the center of attention. As a network, there is a potential for ongoing learning to take place, focusing on research and development in teaching and learning, from which autonomy and continuing professional development can be the natural outcomes.

As the subject is to be understood in the community of truth, the complex patterns of communication emerge. They include sharing observations and interpretations, correcting and complementing each other, torn by conflict in one moment and joined by consensus in the next. The community of truth, far from being linear, static, and hierarchical, is circular, interactive, and dynamic. At its best, the community of truth advances knowledge through conflict, not competition.

This *knowledge-seeking* brings light to understanding, which, in turn expands a sense of autonomy in handling new challenges. Systems inevitably tend to hold individuals accountable, and a nurturing environment can build integration between reform agendas for the accountability needed.

The power of collaborative professional learning can be developed in nurturing environments within schools and as a network. Within this frame of understanding, professional networking can be further developed. The current funding structure inevitably sets different stakeholders—namely policy-makers, school educators, and university educators—in different locations. Respectively, the accountability pressures create tensions relating to self-justifications to the public, particularly when public funding is used.

Recognizing this tension is of primary significance. Stakeholders from various sectors can caution against the pulling of professional energy in different directions and, instead, come to a critical focus on partnership, which is to be built on deep core values of education (see figure 13.2).

This is not to advocate any instant agendas and resolution but to appeal for attention from leaders of different sectors to create this critical space to build an understanding of professional partnership for harmonizing accountability and autonomy. In the process, the role of university academics is to contribute to the building of this learning force and to research the emerging critical focus through scholarly conceptualization.

Conclusion

On the basis of a documentary analysis of official reform agendas, an impact study funded by the government, and a follow-up study from a strategic sam-

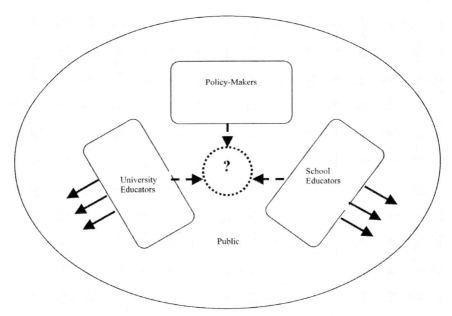

Figure 13.2 Tension between accountability agendas and critical focus

pling of two schools, this chapter has presented an interpretation of tension between accountability and autonomy. It suggested that the ownership of reform initiatives should be reconsidered by strengthening and, if necessary, recasting professional relationships between educational stakeholders of different sectors. A strategic move to the creation of learning space for co-inquiry is arguably the central focus for professional autonomy among educators committed to the end of accountability systems. The agenda is not about teachers being supported to implement reform agendas. Instead, it is through the empowerment of teachers to take charge of their learning while *integrating* reform into their changing practice that the desired changes will take root. Because systems cannot lead to expected impact by themselves, research should be directed toward the development of this learning space that can be created for and by educators.

In conclusion, with the re-orientation of relationships of professional partnership among educational stakeholders, the younger generation can benefit from teachers who are prepared and nurtured continually for autonomous learning in a professional environment, to uphold the goals intended for accountability systems, and be truly liberated from pressures.

References

Advisory Committee on Teacher Education and Qualifications (ACTEQ). (2003). *Toward a learning profession: The teacher competencies framework and the continuing professional development of teachers.* Hong Kong, Government Logistics Department.

Bray, M., & Lee, W. O. (eds.). (2001). *Education and political transition: Themes and experiences in East Asia.* CERC Studies in Comparative Education 1, second edition; Hong Kong: Comparative Education Research Centre, The University of Hong Kong.

Education Commission. (1997). *Quality school education* (Education Commission Report No. 7). Hong Kong: Government of the Hong Kong Special Administrative Region.

Education Commission. (2000). *Reform proposals for the education system in Hong Kong.* Hong Kong: Government of the Hong Kong Special Administrative Region.

Education Commission. (2002). *Progress report on education reform (1).* Hong Kong: Government of the Hong Kong Special Administrative Region. Education Commission website (www.e-c.edu.hk).

Education Commission. (2003). *Progress report on education reform (2).* Hong Kong: Government of the Hong Kong Special Administrative Region. Education Commission website (www.e-c.edu.hk).

Education Commission. (2004). *Progress report on education reform (3).* Hong Kong: Government of the Hong Kong Special Administrative Region. Education Commission website (www.e-c.edu.hk).

Education and Manpower Bureau. (2005). *The new academic structure for senior secondary education and higher education—Action plan for investigating in the future of Hong Kong*. Hong Kong: Government Logistics Department.

Kwo, O. (2007). Towards a learning profession: Understanding induction in a district-based community. In J. Butcher and L. McDonald (Eds.), *Making a difference: Challenges for teachers, teaching, and teacher education*. Sense Publishers: at press.

Levin, B. (2001). *Reforming education: From origins to outcomes*. London: Routledge Falmer.

MacBeath, J., & Clark, B. (2005). *Final report of the impact study on the effectiveness of external school review in Hong Kong in enhancing school improvement through school self-evaluation: Executive summary* (www.emb.gov.hk).

Mehrahtu, T., Crossley, M., & Johnson, D. (eds.). (2000). *Globalisation, educational transformation and societies in transition*. Oxford: Symposium Books.

Palmer, P. J. (1998). *The courage to teach: Exploring the inner landscape of a teacher's life*. San Francisco: Jossey-Bass.

Tung, C. H. (1997). *Building Hong Kong for a new era*. Hong Kong: Government of the Hong Kong Special Administrative Region.

CHAPTER 14

Lessons from Te Kotahitanga for Teacher Education

Russell Bishop
University of Waikato, New Zealand

> Russell Bishop, Ph.D., is a professor and assistant dean of Maori education at the University of Waikato, New Zealand. He is the project director of Te Kotahitanga: Improving the Educational Achievement of Maori students in Mainstream Schools.

ABSTRACT

Te Kotahitanga: Improving the Educational Achievement of Maori students in Mainstream Schools is a funded professional development and research project underway in New Zealand. The approach brings to the fore six challenges from which important lessons for the teaching and learning of minority youth can be derived. These challenges include: the hegemony of the status quo, the primacy of teachers' positioning, the need for evidence, the role of power in knowledge construction, the disconnect between preservice and in-service education, and the fundamental importance of research in the areas of teaching and teaching education. The storying and re-storying of these issues amid a "pedagogy of relations" lead to a powerful form of accountability that countries, leaders, teachers, professors, and families seek.

Introduction

Te Kotahitanga: Improving the Educational Achievement of Maori Students in Mainstream Schools (Bishop, Berryman, Tiakiwai, & Richards, 2003) is a Kau-

---

This chapter was adapted from the keynote address *Messages from Te Kotahitanga for Teacher Education*, presented at the P.R.I.D.E. Workshop held at the National University of Samoa, Apia, Samoa, November 28 to December 2, 2005.

papa Maori research and professional development project that aims to improve the educational achievement of Maori students in mainstream classrooms. While this project addresses Maori students' educational achievement, there are a number of messages from our experiences with working with mainstream teachers and teacher educators that might well be pertinent to situations in which other people find themselves in a post-colonial world.

The Current Educational Context

The major challenge facing education in New Zealand today is that the status quo is one of ongoing social, economic, and political disparities, primarily between the descendents of the British colonizers (Pakeha) and the indigenous Maori people. Maori have higher levels of unemployment; are more likely to be employed in low paying employment; have much higher levels of incarceration, illness, and poverty than do the rest of the population; and are generally under-represented in positive social and economic indicators. These disparities are also reflected at all levels of the education system. While I do not wish to generalize, my understanding is that the New Zealand situation is not unlike other situations elsewhere around the world.

In comparison to majority culture students (in New Zealand these students are primarily of European descent), the overall academic achievement levels of Maori students are low; their rate of suspension from school is three times higher; they are over-represented in special education programs; they enroll in pre-school programs in lower proportions than other groups; they tend to be over-represented in low stream education classes; they are more likely than other students to be found in vocational curriculum streams; they leave school earlier with fewer formal qualifications; and enroll in tertiary education in lower proportions.

Despite the choice provided by Maori medium education in New Zealand, and decades of educational reforms and policies flying under the banners of multiculturalism and biculturalism that have sought to address these problems, for 90% of the Maori students (Matauranga, 2001) who attend mainstream schools, there has been little if any shift in these disparities since they were first statistically identified over forty years ago (Booth & Hunn, 1962).

Six Challenges for Practice and Practitioners

This problematic situation raises a number of challenges for teachers and teacher educators both in New Zealand and overseas.

CHALLENGE #1: THE ESTABLISHED STATUS QUO OF ETHNICALLY BASED EDUCATIONAL DISPARITY

The major challenge that faces educators today is the continuing disparities of outcomes within our education system. This is seen where Maori children and those of other minority groups are consistently over-represented in negative education indicators and under-represented in the positive—as detailed earlier. In terms of qualifications, Maori students in mainstream schools are not achieving at the same levels as other students, and this situation has remained constant for some time. For example, in 1993, 4% of Maori students gained an A or a B Bursary—a secondary school exiting exam that was used in New Zealand prior to 2004 to determine scholarship and university entrance eligibility—and 33% of Maori students left school without qualifications. Yet, some 10 years later, in 2002, 4% of Maori students gained an A or a B Bursary, and 35% of Maori students left school without qualifications. In effect, little changed over that decade despite numerous efforts made by educators.

Similarly, in 1998, 74.1% of candidates gained university entrance, of which 6.1% were Maori (1,247). In 2002, 87.2% of candidates gained university entrance, of which 6.3% were Maori (1,511). In other words, there was an absolute increase in numbers, but a relative decline overall. Exacerbating this situation was that from 1994 to 2003, retention rates for Maori boys to age 16 fell by 12.4% and those for Maori girls by 7.1%. For the same period, retention rates for non-Maori boys fell by 0.7%, whereas the rate for non-Maori girls increased by 1.4%. In addition to these statistics of disparity over time, statistics also show that Maori children are referred to specialist services for behavior problems at far greater rates then other students and make up 47% nationally (this figure is far higher in some regions) of those suspended from school, while only constituting 21% of the school population (Matauranga, 2004).

The ongoing nature of these problems suggests two major implications. The first is that the status quo in New Zealand education has become one where educational disparities are ethnically based, and despite many protestations to the contrary, this has been the case for over 40 years. The second is that despite the best intentions of educators from schools, colleges of education, and policy agencies, currently we do not seem to have a means of systematically addressing these disparities.

The question therefore arises as to how teacher educators are going to assist and educate student teachers to be able to produce equitable outcomes for children of different ethnic, racial, cultural, class, and language groups when they become practicing teachers in the face of these long-term and seemingly immutable disparities. The first thing they need to do, I maintain, is to examine their own discursive positioning, those of their students, and the impact that these perspectives might be having on student achievement. By discursive positioning, I mean how teachers construe the complex historical phenomena experienced by Maori youth and how they stand as educators in the situation. In other words, which sets of ideas and actions (that is, discourses) do educators draw upon to explain their experiences?

CHALLENGE #2: TEACHER POSITIONING

All educators hold a variety of discursive positions with respect to the challenge posed by minority students. In Bishop et al. (2003), we found that teachers tend to draw on three major discourses when explaining their experiences with the education of Maori students: the child and their home, school structures, and relationships. The first two tend to locate the problem outside of the classroom and tend to blame the child and/or their home or the school systems and structures for the seemingly immutable nature of the ongoing disparities. The outcome of teachers' theorizing from within these discourses is that change is seen to be beyond the power of the teacher to act or to produce an effect, that is, to have agency or the freedom to act.

In contrast, the discursive position of relationships tends to promote the agency of the teacher in that it acknowledges that ongoing power imbalances within classrooms create educational disparities, and power imbalances can be changed through changes in pedagogy. Such a position is agentic, as in being one of a change agent, which allows teachers the opportunity to examine how they themselves might participate in the systematic marginalization of Maori students in their own classrooms through their discursive positioning.

To Maori theorists (Bishop, 1996; Smith, 1997), it is clear that unless teachers engage in such considerations of how dominance manifests itself in the lives of Maori students (and their "whānau" or extended family), how the dominant culture maintains control over the various aspects of education, and the part they themselves might play in perpetuating this pattern of domination, albeit unwittingly, they will not understand how they and the way they relate to and interact with Maori student may well affect learning. However, an appreciation of relational dynamics, without an analysis of power balances, can promote professional development that promotes ways of "relating to" and "connecting with" students of other cultures without there being a means whereby teachers can understand, internalize, and work toward changing the power imbalances of which they are a part. In particular, teachers need an opportunity to challenge those power imbalances that are manifested as cultural deficit theorizing in the classroom, which, in turn, support the retention of traditional classroom interaction patterns and that perpetuate marginalization.

To this end, Valencia and Solórzano (1997) traced the origins of deficit thinking, including various manifestations such as intelligence testing, constructs of "at-riskness," and "blaming the victim" (see also McLaren, 2003). More recently, in Shields, Bishop, and Mazawi (2005) we detailed how educators and policy makers continue to pathologize the lived experiences of children through our examination of the experiences of American Navajo, Israeli Bedouin, and New Zealand Maori children's schooling.

In general, we detailed the common practice of attributing school failure to individuals because of their affiliation with a minoritized—i.e., to be ascribed characteristics and treated as a minority regardless of whether the group is a numerical minority or not—group within society by a process termed pathologizing. Pathologizing, according to Shields, Bishop, and Mazawi (2005), is a process where perceived structural-functional, cultural, or epistemological deviation from an assumed normal state is ascribed to another group as a product of power relationships, whereby the less powerful group is deemed to be abnormal in some way. Pathologizing is a mode of colonization used to govern, regulate, manage, marginalize, or minoritize primarily through hegemonic discourses (p. 120).

Pathologizing the lived experiences of children is most often seen in deficit thinking and practices and is a form of power that, as Foucault (1980) explains, works on and through individuals as they take up positions offered to them in discourse and as they become objects of discourse in that discourses provide each of us with a self-narrative that we use to talk and think about our positioning within society. We construct meaning out of our experiences in an interactive social process that locates us within sense-making frameworks that are discourses, or languages in action.

Struggles over the representation of what constitutes appropriate knowledge (struggles that are common to colonized peoples) are struggles over whose meaning or sense will prevail. Meaning does not lie in images as such, nor does it rest entirely with those who interpret what they see; "rather it emerges in the dialogue between those who do the interpreting and the images they perceive" (Ryan, 1999, p. 5) and who has the power to determine the knowledge which is most legitimate. Thus, those who are positioned within the dominant discourse have recourse to a means of framing the ways which "subordinate groups live and respond to their own cultural system and lived experiences" (McLaren, 2003, p. 77) rather than referring to the sense-making of those "othered."

This represents a challenge for educational reformers, teacher educators, and teachers alike in that, as Bruner (1996) identified, it is not just a matter of intervening in part of the system. What is necessary is that we challenge whole discourses and move ourselves beyond our current positionings to alternative discourses that offer educators an opportunity to act as change agents.

The main challenge this understanding poses for teachers and teacher educators is the increasing irrelevance of deficit positioning as a theoretical space from which to develop teaching practice. Yet, in a frequency count of unit ideas commonly used in explaining their experiences with Maori students, a group of teachers (Bishop et al., 2003) drew most commonly upon deficit discourses to explain the educational disparities facing Maori students with consequent feelings of anger and frustration about their lack of agency.

In Te Kotahitanga, we have identified that when teachers locate themselves in these deficit positions, and these are the most common, they blame others for educational disparities, they exhibit feelings of helplessness, and they reject their personal and professional responsibilities and agency. In contrast, when teachers actively reject deficit and blaming explanations, they accept personal and professional responsibility for their part in the learning relationships, they are clear that they have agency in that they are powerful agents of change, they know how and what to do in their classrooms to bring about change, and they report being reinvigorated as teachers. What is problematic for education in general, and that which needs to be addressed specifically by schools and teacher educators alike, is that the majority of teachers position themselves within deficit discourses, thus limiting their agency and, hence, their students' achievement.

Identifying discursive positioning involves teacher education students, staff, and teachers engaging in ongoing opportunities to critically reflect upon evidence of the impact of the positions they hold on student learning. Therefore, such questions as "How do we provide our students/teachers with these opportunities?" are important. Another implication of this understanding is that this reflection will necessarily involve those currently outside of the current reference groups because to continue to talk to a small group of people tends to reinforce the range of discourses open to student teachers. Other discourses are needed, and, therefore, widening the range of discourses open to student teachers is vital.

Ryan (1999) identifies a number of strategies: challenging racist discourses; critically analyzing mass media and contemporary and historical curriculum resources; fostering cultural identities and community relations; and valuing different languages, knowledges, and alternative discourses. One effective means of employing this latter strategy has been used by us in Te Kotahitanga (Bishop et al., 2003) where narratives of the experiences (Connelly & Clandinin, 1990) of a number of Maori students have been used at the commencement of a professional development program with teachers and school leaders to challenge these educators to reflect upon their own positionings vis-à-vis the lived realities of these students and to examine the discourses within which they and the students position themselves. (Note: Other alternative narratives of experience are to be found in New Zealand Maori people's experiences of resistance to colonization through the development of Kaupapa Maori educational initiatives such as Te

Kohanga Reo [Maori medium pre-schools] and Kura Kaupapa Maori [KKM], Maori medium primary schooling]).

The major finding of this aspect of Te Kotahitanga is that education professionals who position themselves within deficit discourses that pathologize the lived experiences of minoritized students are actually disempowering themselves from achieving the very goals that they themselves wish to achieve in terms of their students' academic achievement. Teacher educators, teachers, and student teachers need to be supported and to support one another to accept the primacy of their agency as educational professionals and the responsibility for their actions that such a position entails. On the other hand, not doing so creates problems for themselves and the students they wish to teach.

In order to bring about change in student outcomes, teacher educators should create contexts for learning where the plague of blame is replaced with a culture of agency. Once this has been achieved, teachers are then in an appropriate space to learn how to develop and change their practice through the use of a wide range of evidence and to take responsibility for the changes that are necessary to their teaching practice in response to ongoing formative assessment of student achievement. For example, student teachers will then be able to learn how to set and measure achievement goals for minoritized students and what to do with the information if and when we get it. This latter expectation, of course, raises the issue of how preservice and in-service teachers are going to undertake this activity.

CHALLENGE #3: THE CALL FOR EVIDENCE

Among educators there is an increasing demand that teachers understand how to engage in critical reflection on student learning that is evidence-based rather than assumption-based. That is, there is an expectation that evidence will inform educators' problem-solving in a manner that enables them to change their practice in response to student learning.

The implications of this position for teacher educators is that they need to ascertain if they and their students are able to use data to identify how minoritized students' participation and learning are improving—data such as students' experiences of being minoritized, student participation, absenteeism, suspensions, on-task engagement, and student achievement. Such data can then be used in a formative manner so that appropriate changes can be made to teachers' practice in response to students' schooling experiences and progress with respect to learning.

In their recent research on developing and sustaining a program for the improvement of the teaching of reading to 5 and 6 year olds, Timperley, Philips, and Wiseman (2003) found that when achievement information was used by classroom teachers to inform their teaching practice, those teachers were able to monitor constantly the effectiveness of that practice. When necessary, teachers were then able to adjust their teaching methods to ensure that the learning needs of the child were being addressed. In this way, by using both formative and summative assessment to guide the single objective—improving Maori children's achievement—teachers received timely and regular information on the effect of their efforts, and "successful actions are reinforcing and likely to be repeated . . . practices that are new and unfamiliar will be accepted and retained when they are perceived as increasing one's competence and effectiveness" (p. 130).

In such an approach, one pedagogic style cannot be preferred over another because achievement is the sole criterion for the determination of teaching method. In Timperley et al.'s (2003) study, the data were used to prompt change in teaching practice where it was found that a particular teaching method was not working for a specific child. It therefore became possible for "the main measure of the effectiveness of professional development [to be] the extent to which it results in improved student learning and achievement" (p. 131).

Standardized tests were used in this case and can provide schools with data that are critical to sustaining and maximizing the benefit of the practice, albeit where there is a degree of match between what is being taught and what is being tested. The tests potentially measure children's collective progress and thus the efficacy of pedagogy, the knowledge and skill gaps to which teachers must attend, and the areas of strength exhibited by children.

By way of caution, however, Goldberg and Morrison (2003) warn that these potential benefits do "not come automatically" and that "harmful effects of the tests can offset them, if these are not managed appropriately" (Goldberg & Morrison, 2003, p. 73). They warn that teachers must understand the statistical concepts necessary to interpret test results, must be able to interpret results within the context of other data, and must work in an environment in which such results are taken seriously. They argue that the judicious use of standardized testing is more likely to occur when there exists a strong professional community that examines data with a good mix of curiosity and skepticism. Therefore, it is suggested that such activities are best not undertaken in isolation.

Timperley et al. (2003) also found that schools which were making a difference to children's achievement held regular meetings to focus on teaching strategies for children whose progress was not at the expected rate. These meetings were held with a sense of urgency and were supported by senior teachers working with other teachers in their classrooms to assist them in developing new strategies for these children. School-wide commitment to the urgency and centrality of structured and focused meetings of the professional learning community was also found to be essential.

The Timperley, Phillips, and Wiseman (2003) study identified that when teachers were organized into groups and worked together as a professional learning community, with regular meetings where they considered the evidence of student progress and achievement so as to inform their collective progress, they were able to update their professional knowledge and skills within the context of an organized, school-wide system for improving teaching practices. In addition, teachers' efforts, individually and collectively, "are focused on the goal of improving student learning and achievement and making the school as a whole become a high-performing organization" (p. 132).

Clearly, therefore, the implication for teacher educators is that they need to be creating contexts for learning where their students are able to participate in professional learning communities in which student evidence is the focus of problem-solving conversations among its members. Through this approach, student teachers will learn and practice how to set and measure and re-set achievement goals for minoritized students. Furthermore, they will learn what to do with the information when they get it.

CHALLENGE #4: REALIZATIONS ABOUT LEARNING

There is an increasing realization that learning involves constructing knowledge individually and socially rather then receiving it from others. There is also an increasing realization that knowledge is situational and not gender or culture free. In addition, it is always created and promoted for a specific defined purpose and often these purposes promote the language, culture, and values of those in power.

Teachers retain power and control over what knowledge is legitimate in their classrooms by constructing what Australian educationalist Robert Young (1991) terms the traditional classroom as a learning context for children. Young states: "The [traditional] method [classroom] is one in which teachers objectify learners and reify knowledge, drawing on a body of objectifying knowledge and pedagogy constructed by the behavioral sciences for the former and empiricist and related understandings of knowledge for the latter" (p. 78).

To Young (1991), in the traditional classroom, teachers see their function "as to 'cover' the set curriculum, to achieve sufficient 'control' to make students do this, and to ensure that students achieve a sufficient level of 'mastery' of the set curriculum as revealed by evaluation" (p. 79). The learning context these teachers create aims to promote these outcomes. In these classrooms it is teachers who are "active" and who do most of the "official" talk (classroom language).

Technical mastery of this language and the language of the curriculum (which is generally one and the same thing) are pre-requisites for pupil participation with the official "knowledge" of the classroom.

The learning context that is created in traditional classrooms is such that there is a distinct power difference between teacher and learner that, as Smith (1997, p. 178) suggests, may be reinforced ideologically and spatially. Ideologically, the teacher is seen as the "font of all knowledge," the students—in Locke's terms—are seen as the *tabula rasa*, the empty slate; the teacher is the 'neutral' and objective arbiter and transmitter of knowledge. Knowledge, however, is selected by the teacher, guided by curriculum documents and possibly texts that are created from within and by the dominant discourse. Furthermore, this knowledge is founded in colonial and neo-colonial contexts, from outside the experiences and interests of the very people it is purported to educate. Far from being neutral, these documents actively reproduce the cultural and social hegemony of the dominant groups at the expense of marginalized groups.

The spatial manifestation of difference in traditional classrooms can be seen in "the furniture arrangements within the classroom, in the organisation of staff meetings, and by holding assemblies with teachers sitting on the stage and so forth" (Smith, 1997, p. 179). Children who are unable or who do not want to participate in this pattern are marginalized and fail. Teachers will then explain the children's lack of participation in terms of pupil inabilities, disabilities, dysfunctions, or deficiencies, rather than considering that it may well be the very structure of the classroom that works against the creation of a relationship that will promote satisfactory participation by students.

In contrast, what Young (1991) terms a discursive classroom is one where new images and their constituent metaphors are able to be present to inform and guide the development of educational principles and pedagogies in order to help create power-sharing relationships and classroom interaction patterns within which young Maori and other minoritized peoples can successfully participate and engage in learning.

Discursive classrooms that are created by teachers who are working within Kaupapa Maori reform projects, such as Te Kotahitanga, suggest new approaches to interpersonal and group interactions that have the potential to move Aotearoa/New Zealand educational experiences for many children of diverse cultural backgrounds from the negative to the positive. Te Kotahitanga practices suggest that where the images and the metaphors we use to express these images are holistic, interactional, and focus on power-sharing relationships, the resultant classroom practices and educational experiences for children of other than the dominant group will be entirely different.

New metaphors are needed in teaching and teacher education that are holistic and flexible and able to be determined by or understood within the cultural contexts that have meaning to the lives of the many young people of diverse backgrounds who attend modern schools wherever they may be situated in the world. Teaching and learning strategies which flow from these metaphors need to be flexible and allow the diverse voices of young people primacy. In such a pedagogy, the participants in the learning interaction become involved in the process of collaboration, in the process of mutual story-telling and re-storying (Connelly & Clandinin, 1990), so that a relationship can emerge in which *both* stories are heard, or indeed a process where a new story is created by all the participants.

Such a pedagogy addresses Maori people's concerns about current traditional pedagogic practices being fundamentally monocultural and epistemologically racist. This new pedagogy recognizes that all people who are involved in the learning and teaching process are participants who have meaningful experiences, valid concerns, and legitimate questions.

The implications of this understanding for teaching and teacher education is that there is an increasing realization that teachers have the agency to construct contexts wherein students are able to bring their cultural experiences to the learning conversation, despite the teacher not knowing about these experiences and ways of making sense of the world. At the same time, teacher educators need to create learning contexts where their student teachers can experience such relationships and interactions.

CHALLENGE #5: RELATIONSHIP BETWEEN PRESERVICE AND IN-SERVICE EDUCATION

There is an increasing demand from various sectors of the profession for increased relevance between preservice education and in-service education, professional development, teaching practice, and research. This is further exacerbated by international research that identifies that there is little if any linkage between preservice teacher education and in-service practice and by the perceived hierarchies within the education sector (Cochran-Smith & Zeichner, 2005).

From our experiences in Te Kotahitanga, an added problem is that teacher educators, teacher support staff, school teachers, and educational researchers tend to suggest to us that what they are doing is sufficient, necessary, and adequate, in contrast to the functioning of those people in every other sector. In other words, what is happening in their patch is fine; it is all those other people who are not doing a good enough job. Similar findings have been made by Prochnow and Kearney (2002) in a study they conducted about the effect of suspensions on student learning. They found that all the groups of people involved with the students tended to blame others for the problems the students faced and were less likely to implicate themselves in the problem identification process.

To make matters worse, these notions are supported by the process that teacher educators have devised to review their programs—that is, by peer review. These reviews do not usually include their client groups, or if they do, it is in a prescribed manner thus limiting the type of critique that would be useful in reforming teacher education programs so that their graduates would be able to address the learning needs of minoritized peoples.

Other problems that are voiced about teacher education by those in other sectors include the increasing concern about the frailty of the "silo" model for preparing preservice teachers and the continued criticism of tertiary teacher education providers. These criticisms come from teacher education graduates, the education profession, the public, and the media, and are aired in media that is not part of the formal review process. A means of addressing these criticisms is needed urgently.

One example of the problematic response to criticism is found in a recent survey of teacher preparedness that was conducted by the Education Review Office (ERO) (2004). The report, which was critical of the preparation of beginning secondary and primary teachers, was met with criticism by teacher educators and researchers alike. The criticism was leveled at the process whereby this finding was attained, rather than the finding itself, or at least the problems that the survey was indicating could be present. What is of concern is that this reaction did not reenergize the debate, but rather killed the conversation, despite many teachers and schools voicing concern. Yet, recent observations of 360 teachers in Te Kotahitanga, 60% of whom had been to teacher education institutions in the past five years, showed that while they wanted to teach in ways they had learned while at their college of education, they were in fact teaching in a very traditional manner in their first year of teaching.

When surveyed, they stated that they were keen to implement a wide and effective range of interaction types: to actively engage their students in the lessons, use the prior knowledge of students, use group learning processes, provide academic feedback, involve students in planning lessons, demonstrate their high expectations, stimulate critical questioning, recognize the culture of students, and so on. However, detailed, measured observations of their classrooms showed that 86% of their interactions were of a traditional nature where they were engaged in the transmission of pre-determined knowledge, monitoring to see if this knowledge had been passed on and giving behavioral feedback in order to control the class. Only 14% of their classroom interactions allowed them an opportunity to create learning relationships to which they initially aspired.

In short, despite their aspirations to the contrary, the dominant classroom interaction remained an active teacher and passive students. This might signal the pervasiveness of transmission education, in which case we could blame the schools and their insistence on transmitting a pre-set curriculum. However, it might also indicate the lack of student preparedness and the reliance upon the school for practical training, in which case teacher educators could well take notice of the survey and Te Kotahitanga results as a warning that their graduates may be facing problems of classroom implementation of interactive approaches. In other words, these findings might signal the need for preservice teachers to integrate the theory and practice of teaching and learning (using evidence of behavior as teachers and student achievement for formative purposes) in a systematic manner so that they can practice what they learn.

One way this might happen is for preservice teachers to receive objective analysis and feedback of their classroom interactions in an ongoing manner upon which they critically reflect in a collaborative, problem-solving setting. This means that preservice teachers will need to learn to use evidence of student participation and achievement to inform their practice, (to change classroom interaction patterns for instance) and the relationship between teacher education institutions and schools will need to change dramatically.

CHALLENGE #6: THE CHALLENGE OF RESEARCH

The recent Performance Based Research Fund (PBRF) report (Alcorn et al., 2004) states that 75% of staff involved in teaching degree level courses in education are not involved in research. Further, the area with the lowest quality of research and the lowest assessed research performance is Teacher Education. Therefore, if change is necessary to address disparities; if research is our most common way of informing and promoting change through the systematic production of evidence to inform our practice; and if teacher educators are not involved in research, what mechanism are they using to inform their practice? This may mean that despite their avowed aspirations to address what Fullan (2005) terms the moral dimension of education, that is, the reduction of disparities, teacher educators may not have a means of addressing the status quo that is maintaining the disparities that they say they want to reduce.

Conclusion

This chapter has suggested that reducing the seemingly immutable educational disparities in the education system in Aotearoa/New Zealand is in fact possible and the answer lies in a critical examination of the discourses within which teachers position themselves. Commonly, discourses that promote deficit notions that in turn pathologize the lived experiences of Maori students and the schooling systems limit the agency of teachers to make the difference for their students to which, ironically, they aspire. On the other hand, positioning within change-agent discourses allows teachers to take responsibility for their student's learning, to reflect upon evidence of this learning so as to revise their teaching approaches, and to enjoy teaching.

When teachers are (re)positioned within relational discourses and promote what Sidorkin (2002) calls a Pedagogy of Relations, teachers are able to address power imbalances within their classrooms, within their schools, and between the representatives of the various sectors of education who are currently critical of each other. In addition, research becomes part of the everyday lives of teachers and proves its usefulness in both formative and summative manners. In the midst of complex situations and discourses formed around the nexus of relationship, powerful accountability will arise.

Above all, in terms of student achievement, this chapter suggests that the classroom should be a place where young people's sense-making processes (culture with a small c) are incorporated and enhanced, where the existing knowledges of young people are seen as "acceptable" and "official," in such a way that their stories provide the learning base from whence they can branch out into new fields of knowledge. In this process, the teacher interacts with students in such a way (storying and re-storying) that new knowledge is co-created. Such a classroom will generate totally different interaction patterns and educational outcomes from a classroom where knowledge is seen as simply something that the teacher makes sense of and then passes onto students.

References

Alcorn, N., Bishop, R., Cardno, C., Crooks, T., Fairbairn-Dunlop, P., Hattie, J., Jones, A., Kane, R., O'Brier, & Stevenson, J. (2004). Enhancing education research in New Zealand: Experiences and recommendations from the PBRF education peer review panel. *New Zealand Journal of Education Studies, 29*(2), 275–302.

Bishop, R. (1996). *Collaborative research stories: Whakawhanaungatanga*. Palmerston North: Dunmore Press.

Bishop, R., Berryman, M., Tiakiwai, S., & Richards, C. (2003). *Te Kotahitanga: The experiences of year 9 and 10 Maori students in mainstream classrooms*. Wellington: Ministry of Education.

Booth, J. M., & Hunn, J. K., New Zealand Department of Maori Affairs. (1962). *Integration of Maori and Pakeha*. Wellington.

Bruner, J. (1996). *The culture of education*. Cambridge, MA: Harvard University Press.

Cochran-Smith, M., & Zeichner, K. M. (2005). *Studying teacher education: The report of the AERA panel on research and teacher education*. Lawrence Mahwah: Erlbaum Associates Inc.

Connelly, M., & Clandinin, J. (1990). Stories of experience and narrative inquiry. *Educational Researcher.* June–July, pp. 2–14.
Education Review Office (2004). *The quality of Year 2 beginning teachers.* Education Evaluation Reports: Wellington.
Foucault, M. (1980). *Power/knowledge: Selected interviews and other writings.* (ed. Colin Gordon). New York: Pantheon.
Fullan, M. (2005). Resiliency and sustainability. *School Administrator, February/62*(2), pp. 16–18.
Goldberg, B., & Morrison, D. M. (2003). Con-nect: Purpose, accountability, and school leadership. In J. Murphy & A. Datnow (Eds.), *Leadership lessons from comprehensive school reforms.* Thousand Oaks, CA: Corwin Press.
Matauranga, N. H., New Zealand Ministry of Education. (2004). Annual Report on Maori Education 1999/2000 and directions for 2004. www.minedu.govt.nz/web/document/ (retrieved on May 3, 2006).
McLaren, P. (2003). *Life in schools: An introduction to critical pedagogy in the foundations of education.* 4th Edition. Boston, MA: Pearson Education.
Prochnow, J., & Kearney, A. (2002). *Barriers to including students with difficult behaviour: What are we really saying?* Paper presented to the New Zealand Association for Research in Education, Palmerston North, 5–7 December.
Ryan, J. (1999). *Race and ethnicity in multi-ethnic schools: A critical case study.* Clevedon: Multilingual Matters.
Shields, C. M., Bishop, R., & Mazawi, A. E. (2005). *Pathologizing practices: The impact of deficit thinking on education.* New York: Peter Lang.
Sidorkin, A. M. (2002). *Learning relations.* New York: Peter Lang.
Smith, G. H. (1997). *Kaupapa Maori as transformative praxis.* Unpublished Ph.D. thesis: University of Auckland.
Timperley, H., Phillips, G., & Wiseman, J. (2003). *The sustainability of professional development in literacy, parts one and two.* Auckland: University of Auckland.
Valencia, R. R., & Solórzano, D. G. (1997). Contemporary deficit thinking. In R. R. Valencia (Ed.), *The evolution of deficit thinking.* Washington, DC: Falmer Press.
Young, R. (1991). *Critical theory and classroom talk.* Clevedon: Multilingual Matters.

Summary and Implications

Neil J. Liss
Willamette University

What are the "underpinnings of powerful accountability systems" as shown through these three chapters? Will discovery of those underpinnings become the loci for an even stricter form of accountability? These studies show the complex density of standardizing even fluid and open-accessed inquiry. Considering the following implications of the studies here and through a discussion of potential extensions, I hope, will generate commentary on what this means for educational research.

Experience First

There is something inescapable in the findings about educational excellence, as defined by the systems of standardized accountability now being legitimized across the globe, that seems to come *after* the formation of learning communities. By constructing a set of practices around students' engagement of their social world, and through building vibrant relationships that allow for collective thought and action, schools can foster significant results. Ways of sustaining relationships—Bishop's "discursive" pedagogies; the supporting/nurturing "learning spaces" seen in Kwo's research; the autonomous curriculum interpreting for the sake of student/society that Estola, Lauriala, Nissilä, and Syrjälä project as a pedagogical necessity—should occupy theory and practice. This dialectic tethers the social context of the school to the curriculum. The relationships then can be seen as having nurtured demonstrable examples of social justice, democratic humanism and concern for others, and social networks of care rather than calculating materialism. (After all, Estola et al. show that such concern, rather than inhibit economic growth, seems to raise it: Finland being one of the richest twenty nations, with a relatively small income gap.)

Educators, in cultivating relationships, stake ownership "not to external standards, but with respect to their influence on students" (from Estola et al.) and build upon sustained human relatedness. Expectations of rigorous thought can thus be seen to "produce" strong statistical effects, even though disaggregated abstruse "facts" are not the ultimate goal for learning (in reference to the PISA "miracle" in Estola, and to the high-achieving schools documented in Kwo). The numbers, per se, provide documentation on the sustained power of accepting human contingency.

However, it would be rash to make these practices a new end for education. If outcomes *are* to be measured, we need to remember the post hoc nature of measuring. The perils of accountability force into concrete what is otherwise fluid, flexible, and radically particular. We are driven to accept the correlations and inferences, in effect, legitimizing the practice of accountability. Without the ability to confront the forces behind why some forms of accountability reign supreme, even qualitative narratives of accountability get sucked into a quantification of education. As ends in itself, forms of evidentiary "proof" of educational success can usurp the collective effort of nurturance, support, and deliberative meaning-making.

Change Is Happening

Each of these studies took place at a time of, if not rapid change to their sociocultural foundations, a growing *sense of* foundational change. And in each, education served as a means toward this awareness, explicitly positioning the students and teachers to effect more accountable change. Collective consciousness about social structures helps to bring out alternative and pluralistic approaches to knowing and acting. For the Finnish teachers, their preparatory work in education schools grounds their pedagogy in inquiry. Personal experiences of the most relevant issues help organize their curriculum choices. They come into classrooms having internalized tacit accountability to influence change, not to passively accept it.

As Kwo gives insight, reflective thought institutionalized across a school leads toward responsible decision-making. Here, in the context of new paradigms of social structure (stemming from the 1997 "reversion of sovereignty"), school leaders support institutionalized reflective thought because it has produced—again, accountability as after affect—positive measures. As teachers accept their leaders' autonomy in implementing measures mandated by the government, leaders grant teachers flexibility in interpreting them. This "mutuality" pokes the educational enterprise outside of its narrow confines of statistical accountability.

Liberty

All three chapters also speak of the need for iterations of teacher discretion through interpreting and devising curriculum, through individual reflective practice and ongoing dialogue, and in challenging the top-down miasma that clogs learning. Teachers achieve "authenticity" and "agentic" action, as Bishop tells us, less in formal role-taking than in mindful, though certainly rigorous and disciplined, *decision-making*. Their decisions, emergent when real-time experiences are integrated into classrooms, are based upon the burgeoning relationships that support a direct confrontation with the very foundations that extenuate the decision-making. Because this autonomy leads to accepting students' emic understandings, education becomes the process for connecting their grasping of the world to the expected social understandings necessitated *by* their world. Loosened from the backdrop of articulated discourse practices through research pedagogy, students confront their own autonomy as persons within a community of ethics and responsibilities.

Autonomy cannot be given, in the sense of a privilege that can be taken away. The freedom referenced here relates to a kind of *phronesis*, or ethical judgment. This wisdom relies on personal, practical knowledge explicitly tapped, as well as the tacit dimension to experience that comes to life in full engagement. *Phronesis* is needed to locate educational practice within the fluid contexts of relationships, social change, and institutional expectations. No formula exists for this. Teachers must commit to underlying principles that weigh heavily the contexts which frame learning. Aligned with those bigger questions that surely define who we are, these principles, perhaps best thought of as the broad, holistic aspirations to which any society holds itself, must be the final accountability metric.

Empirics Matter

Referencing data is paramount. The hard stuff of numbers, *among* a slew of various kinds of measuring, helps teachers confront the gap evinced by overly specified outcomes given to them by the larger community. They can aid in teachers moving from "assumptions" about what is happening in their classrooms to "evidence" of what kinds of learning actually are present. And yet, each of these studies shows how educators should hold documentation of every kind up to scrutiny, digging beneath a shallow acceptance of raw empirical evidence. This "reminder to reflect," as Kwo puts it, is the "beginning" and not the end of accountability. Educators must do the thought-work necessary to contextualize all data gleaned from their practice.

By acknowledging the statistical "proof" of their PISA scores, Estola et. al. show, perhaps unfortunately, its manipulative uses. When they work for your practice, exonerate them, promote them, refer to them as a "proof" of your school's academic "success story." I doubt a sense of fatalism in these researchers as much as I empathize with how the educational field has internalized statistical forms of accountability. The "good" schools must rely on these simple, yet veiling forms of accountability to begin a conversation about responsible education. Without this starting point, schools lack the institutional traction to bring the community into the dialogue about what needs be done.

Where/Everywhere

Perhaps this is but a semantic difference, but each of these studies highlights a separate node in the educational web. For the Finns, reflective action research must take place in *schools of education* (Bishop also emphasizes this), so that pre-service teachers bring a refined linkage of theory to practice into their classrooms. For Kwo, the *school* becomes a "space" for a broader learning community (which Estola et. al. recognize as well), to interpolate the "private aspect of reflectivity" with the "public aspects of desirable performance." And for Bishop, the *classroom* itself becomes the site for the whole system change (as both Estola et. al. and Kwo recommend) as "discursive relationships" go to work reassembling knowledge-structures.

The distinctions fall away if we look at education in its broadest terms. Teachers learn in the process of teaching, exploring with their students. Leaders cannot dwell in calculating efficacy, for then they are not supporting teacher autonomy. And students can learn to examine "their own discursive positioning" only if their teachers and school leaders are willing to do so. Each level incorporates learning as its texture, eschewing problem-solving, narrowly construed as discrete, short-term instrumental rationality, as the modus of educational philosophy.

Extensions

Education cannot do away with stable kinds of inquiry, nor with performance-based accountability. These can be welcomed, especially if agreement is reached on them through the exact kinds of intelligent thought and dialogical processes highlighted in the three chapters under discussion. The danger lies when these inquiry or accountability orientations become their own rewards, too stable and too solidly oriented to specific outcomes that they are severed from the living flow of human relatedness. What needs attention, then, are those underpinnings of society that provide stasis amid the flow. This calls educators to form a philosophy, and not just of education. We need to grapple with the worldview we inhabit, for our aspirations about our society life that provide the lens through which we construct educational goals. These studies only tentatively point us toward how education relationships partake in the struggle to decide the good society.

But making inquiry the spark of education is different from making the process of deciding (or relationships or accountability) its endgame. Live inquiry, what Schwab (1982) called "eclectic arts" of acting in the world, defers meaning. Reflection generates ideas from experience; we can observe and know them. We can even *feel* them. But live inquiry requires us not to stop with any particular, useful conception, but to plow back into that self-evident moment of knowledge. The constant rigorous process of reclaiming those rooted, deeply held principles keeps education connected to the unspoken but lived social good. Because it may look different at different times—our local time-worn truths prove fallible—live inquiry is a continual process of critical thought and engagement, grappling with the poignancy of self-understanding. Relationships form, accountability happens, but not passively. Concerted effort on research requires decisive human action.

Bishop's reference of Foucault (1980) aptly positions this section of the *Yearbook*. Knowledge cannot be separate from power; and power structures society such that the meanings we make, though subjective and personal, come about through positions offered to us in discourse. Put another way, we step into meanings as determined by the matrices of power. Transparency relates to how the special structures of consciousness—hardly free-form subjectivity—can keep hidden our role in bringing about those structures. Decisions can be less autonomous than automatic.

Accountability in education reflects this hiding, this automaticity. Educators grasp for overarching ways of setting their educational practices against the backdrop of the ever more-expectant world. As this complexity begets critical questions for students, education lessens the rote acceptance of static forms of participation. Thus the focus on the socio-cultural context ingrains in students a proclivity toward ethical participation in society. Putting this kind of human face on education, though, is difficult. We must accept students' often unintelligible, illogical, and perhaps seemingly irrational efforts at understanding what we expect of them. Students, too, must face this difficulty, by taking responsibility for these expectations. "What" we expect can give way, in inquiry, to "why we expect that." Ignorance of either saps the transformative power of education. It then deals only in mindless finalities and not in living thought.

What these chapters call for as well is something akin to respect for silence.

Here is that bare moment, naked of meaning, from which to push off into the critical space. Teachers need the space and time to come to grips with changing expectations that come from changing social conditions; students need that briefest of pause to begin the act of finding their voice among the symphony of meanings thrown at them. The human condition demands meaning in experience, even though answering that demand risks the flourishing in that experience. There are no abstractions to people; everyone lives fully, even though we may never completely understand each other's particular experience. But we can make sense of whether people learn to meet the expectations we collectively hold out to them. When accountability is limited to just *that* sense and not the underpinnings of quietitude that impassion inquiry, we risk dehumanizing education.

The kinds of evaluation we need of this education—not only summative and quantitative, but formative and nuanced which feeds important information into the instructional loop, rather than saving problems to the end—are generative. Education is renewed as more participants take the pause needed to dwell in this flexible, open-ended process of deciding, acting. How to shape that future activism speaks to the kinds of accountability society should expect from its educators. Loss of teacher autonomy can strangle student *phronesis*. Implementation of abstract and hollow goals of education (cut off from the flow of changing life) standardizes social participation. Compliance becomes a social duty, formalizing the acceptance of the transparent truth of top-down decisions and their systems of accountability. Voice becomes echo; silence, mimesis.

The individual will always be held accountable to the larger world; we live in what we make of it. Human praxis comes in how we make sense out of that participation. Education should expect the kind of critical distancing from what "is" explored in these chapters—one that focuses student awareness on human plurality and toward responsible judgment. Accountability must meet claims for this kind of ethical and difficult context.

> Compare the silent rose of the sun
> And rain, the blood-rose living in its smell,
> With this paper, this dust.
> That states the point.
> —Wallace Stevens, "Extracts from the Academy of Fine Ideas"

References

Foucault, M. (1980). *Power/knowledge; selected interviews and other writings.* Colin Gordon (Ed.), New York: Pantheon.

Schwab, J. J. (1982). "The Practical: A language for curriculum" in *Science, curriculum and liberal education.* Chicago: University of Chicago Press.

Afterword

Cheryl J. Craig
University of Houston

Louise F. Deretchin
Houston A+ Challenge

Our editorial wonders about who would respond to the call for the *International Research on the Impact of Accountability Systems: Teacher Education Yearbook XV* and what authors and respondents would have to say about accountability systems have been answered. A chorus of voices has risen up from locations dotted across the U.S. and around the world. A plethora of perspectives and positions have been presented. Authors have offered amplified accounts of accountability through rendering students', teachers', teacher educators', minority populations', and international points of view, all of which were informed by philosophy, history, and human experience.

Wheatley, Kelchtermans, and Flinders kicked off this collected volume of essays by focusing on the underside of humanity's long fascination with numerical accounts (Wheatley), the rise of accountability/performativity in the international policy arena (Kelchtermans), and the uneasy relationship between standards and standardization embedded in the U.S.'s No Child Left Behind Act (NCLB) (Flinders). Their essays were followed by those authored by Fletcher, Strong, and Villar, Lee, Klug, Snow-Gerono, and A. Ganesh.

In Fletcher, Strong, and Villar, we discover the vast complexities that come into play when attempting to connect a new teacher support program as an extension of the teacher education arena to students' reading achievement in the K–12 schooling domain, which understandably is a stretch with innumerable, extenuating factors strung in between. Also, readers come to know the social, cultural, and academic contexts of educational systems underlying international comparison studies (Lee), the devastating effects of NCLB on historically underserved Native American students (Klug), and the impact of NCLB on preservice and in-service teachers' working conditions and pedagogy (Snow-Gerono and A. Ganesh).

The next group of essays we encountered were authored by Kosnik, T. Ganesh, and Finnell-Gudwein. Kosnik chronicled her personal awakening as a Canadian to how accountability unavoidably shapes the professional lives of American teacher educators (in addition to teachers and students); T. Ganesh conveyed teachers' real stories of experience of accountability when their cover stories were peeled back; and Finnell-Gudwein employed literature reflecting both liberal and conservative traditions to argue that the NCLB Act is non-democratic.

In the last set of essays, three international chapters were featured: the first from Finland (Estola, Lauriala, Nissälä, & Syrjälä), the second from Hong Kong (Kwo), and the third from New Zealand (Bishop). Estola, Lauriala, Nissälä, and Syrjälä centered on Finland's world class PISA (Program for International Students Assessment) results and informed us that the educational success of students in their country is not only about an educational system that promotes the autonomy, professionalism, and reflectivity of its teachers (supported by a high quality teacher education program), but about an expansive social net provided by every facet of society ranging from early childcare parental leave to close attention paid to citizens of all ages who are in need. Kwo mapped Hong Kong's return to China and the necessary tensions between autonomy and accountability, and the development of a shared public space emerging in the post-colonial policy environment after "the reversion of sovereignty." Bishop focused on the education of Mäori students in New Zealand and illuminated how that country is confronting historical inequities and hegemonies also in post-colonial times. In all three instances, the sense of learning on a continuum emerged. What also was evident was support for the conditions of learning, which included, as we particularly note in Bishop's case (chapter 14), a willingness to attend to deeply engrained psychological and contextual matters that require unlearning as well.

Not only did chapter authors provide viewpoints, so did *Yearbook* respondents who spoke from the authority of over 150 years of school-based experience as teachers (Kelley, Reid, Curtis, Venable, Martindell, & Hamacher), from troubling high-stakes accountability test results routinely used in statistically indefensible ways (McCormack, McDonald, T. Ganesh, & Foster) and from contrasting policymakers' original aims underpinning the NCLB Act to artists' initial intentions underlying their masterpieces (Kahn, Lee, Markello, Mullins, & A. Ganesh). And respondent Liss reminded readers that accountability inherently involves an exchange of power and that the best we can hope for is active inquiry and transparency as people work toward visions of accountability framed by mutuality and power-sharing. Such a positioning would ward against

AFTERWORD 249

accountability becoming humanly lived as a unidirectional arrow emanating from those with the most power and pointed at those with the least power.

Throughout this collected volume, many noteworthy connections emerged across chapters and divisions. We remind readers of but a few. Wheatley and Roger's warnings about the uses and abuses of measurement in chapter 1 were echoed in chapters authored by Fletcher, Strong, and Villar (chapter 4), Lee (chapter 5), and T. Ganesh (chapter 10), and became reverberated in McCormack, McDonald, T. Ganesh, and Foster's remarks in division 2; Kahn, Lee, Markello, Mullins, and A. Ganesh's comments in division 3; and Liss' analysis and synthesis in division 4. Flinders' lessons in chapter 3 found company in Bishop's lessons in chapter 14. Lee's international comparisons in chapter 5 gave way to Estola, Lauriala, Nissäla, and Syrjälä's chapter 12. Kelley, Reid, Curtis, Venable, Martindell, and Hamacher's division 1 commentary foreshadowed Snow-Gerono's, A. Ganesh's, Kosnik's, and T. Ganesh's essays in chapters 7, 8, 9, and 10, respectively. Finnell-Gudwein's analysis in chapter 11 fit with Flinders' account in chapter 3. Kelchtermans in chapter 2 and Kwo in chapter 13 had critical dimensions of their analyses—originating from two different continents, contexts, and cultures—in common. To these resonances, we highlight similarities between Klug's essay in chapter 6 that focused on the plight of Native American students and Bishop's concern in chapter 14 for the treatment of a different underserved, indigenous people, the Māori, in New Zealand.

At the same time, significant dissonances surfaced. The attentiveness of Hong Kong's policymakers to the possibility of the accountability system overburdening teachers and interfering with their teaching of students, so evident in Kwo's chapter 13, was utterly absent in the U.S. accounts offered by Flinders (chapter 3), Klug (chapter 6), Snow-Gerono (chapter 7), A. Ganesh (chapter 8), Kosnik (chapter 9), T. Ganesh (chapter 10), and Finnell-Gudwien (chapter 11). Also, New Zealand's attempt to make amends for its historically grievous policies and practices, to which we are introduced in Bishop's chapter 14, appears to be an idea that has been lost on U.S. policymakers, as Klug's chapter 6 sadly makes clear. The same can be said for policymakers' provisions for the poor and the needy in Finland as Estola, Lauriala, Nissäla, and Syrjälä point out, and the seeming absence of policymakers' concern in the U.S.—even for the very young, as A. Ganesh's essay in chapter 8 points out.

Despite the similarities and differences evident in the diverse accounts in this *International Research on the Impact of Accountability Systems: Teacher Education Yearbook XV*, one commonality, however, traversed all accounts. No authors or respondents argued vehemently against accountability. Rather, the concerns expressed had to do with how accountability systems became enacted. In other words, humans ceding authority to inanimate—some went as far as to say inhumane—systems was the core problem. Another central issue was measurement conceived as both the means and ends of accountability—with punitive action taking precedence and generative improvement falling by the wayside. Such an approach was seen as robbing people and situations of their particularities and complexities, indeed, of the richness that distinguishes human meaning-making in context, as Kelchtermans in chapter 2 so eloquently stated.

At the same time, as the international essays in this volume exposed serious weaknesses in the U.S. accountability system, they informed us that the possibility of productive accountability systems exists. We learn, however, that these systems do more than merely graze/brush the educational surface. They run much deeper than successive waves of policymakers depositing their ideological-laden baggage on the educational scene. These strong systems appear deeply embedded in the wills of people and fixed in the psyches of the nations who embrace them. These responsible systems strike at the heart of equity while simultaneously seeking excellence. They do so without being time-consuming, disruptive, hierarchical, rhetorical, or burdensome. They respect human autonomy while reaching the core of educational matters of national and international significance. Such responsive systems appear open to scrutiny and to being internally adjusted to ensure that they are serving people and advancing teaching and learning and not accomplishing the contrary. The ever-present possibility of unintended consequences appears to be humanly held in check.

As we conclude this volume, major questions remain. Once again, we present only a few. At the meta-level, we ponder how the U.S. accountability system can recover its democratic roots given that proponents of both the educational left and the educational right claim that the country's democratic history is being/has been lost, as Flinders (chapter 3) and Finnell-Gudwien (chapter 11) suggest. At the same time, we reflect on how aspects of others' accountability systems such as the one lived in Finland (as presented by Estola et al. in chapter 12), the one lived in Hong Kong (as reported by Kwo in chapter 13), and the one lived in New Zealand (as chronicled by Bishop in chapter 14) could inform how accountability could be enacted in highly productive ways in the U.S. We concurrently wonder what would happen if academics like Kelchtermans in Belgium (chapter 2) headed up policy think tanks in the U.S.

Then, at the micro level, our thoughts travel to the generation of students, teachers, and teacher educators for whom statistical accounts have judged their worth, but not contributed to their growth. How can the growth orientation be embraced by individuals without victimizing those who have been the victims of capricious policymaking? And how does the U.S. deal with the rapidly swelling populations of not one, but several underserved groups?

Finally, as editors, we recognize that additional research is needed. One such area concerns the unintended consequences of high-stakes testing (as the vast majority of chapters in this volume make resoundingly clear); a second is inquiry into the conditions of teaching and learning as experienced by both teachers and students (as many chapters indicated), particularly those representing underserved populations (as certain chapters emphasized); and a third is the study of how the dialectics of autonomy and accountability emerge as vital parts of professionalism in preservice education and in-service teacher development (as some chapters stressed). Lastly, we believe that the final yearbook we compile during our editorship, *Teacher Education Yearbook XX*—to be released in 2011, should be dedicated to a theme similar to this one in order to show how history continued to unravel and what transpired in the post-election period to the NCLB Act of 2001.